Diogenes Station

A Perspective on Transformational Leadership
Thirty Years Later

or

Sex Is Not Our Policy

William G. Hanne
Colonel (U.S. Army Ret.)

Printed in the United States of America

First Printing September 2016

ISBN 978-1-944787-31-8 Paperback

ISBN 978-1-945177-12-5 Hardcover

Published by: Book Services
www.BookServices.us

Contents

▦ = Photos & Maps

Preface

The opportunity for a commissioned officer to serve as a commander is a rare occurrence, given the size of the U.S. Army and the number of officers eligible for such a position. Given also the pyramidal structure of the Army, it is obvious that as one goes up in rank, the opportunities for command become fewer. So to be selected for command as a colonel is both an honor and a privilege.

The process of selecting an individual for command has changed over the years, from political favoritism, to being in the right place at the right time, to the current selection process. Selection boards of senior officers are convened annually. These boards determine who is eligible for command at the lieutenant colonel and colonel ranks, based on the written records (effectiveness reports). The command selectee rosters are turned over to the respective branch—e.g., Infantry, Artillery, Military Intelligence, Signal, etc.—for the determination of where the selected individuals will be assigned.

Some O-6 commands[1] are very visible, such as the commander of the Old Guard, responsible for the Tomb of the Unknown Soldier and for ceremonies in the D.C. area. Other O-6 commands are hidden in the bureaucracy of "the system," some deliberately, so that little or no attention is drawn to them because of the tasks involved.

Field Station Sinop was one of the latter. Located on the north coast of Turkey, overlooking the bay where the Crimean War began in 1853 with the sinking of the Ottoman fleet by the Russian navy, Diogenes Station was a joint (i.e., a service besides

---

1. O-6 is jargon for the rank of colonel—it could be viewed as the sixth rank in a structure which starts with O-1, Second Lieutenant, and ends in O-10, a four-star General.

the U.S. Army was on-site), as well as a combined (i.e., the armed forces of another nation were on-site), NATO installation focused on electronic communications.

Field Station Sinop had U.S. Army and U.S. Navy personnel present, plus a representative of the U.S. Air Force and Turkish armed forces. Taken together, Diogenes Station, as the field station was frequently called, was a combined/joint installation under the responsibility of a Turkish colonel. (It had yet another title: "TUSLOG Det 4."[2]) The site was in Turkey, and the Status of Forces Agreement (SOFA) between the United States and the Republic of Turkey made Colonel Senal the overall installation commander. (In this instance, the SOFA was the Defense and Economic Cooperative Agreement (DECA) that spelled out responsibilities and limits of authority.) Each of the organizations—the Turkish Army, the U.S. Navy, and the U.S. Army—had its own commander who was responsible to his chain of command for what got done or did not get done, which made life interesting, to say the least.

Because Field Station Sinop was an isolated post, family members were precluded from accompanying the service member at governmental expense. If a family member—in the vernacular of the "Old Army," the term was "dependent," but whoever referred to my wife as a "dependent" took his or her life into their own hands—chose to accompany the service member at their own expense, that was a different story. The site offered nothing in the way of family accommodations, medical or dental care, commissary, PX, or any other dependent-related facilities.

In contrast, in "Today's Army" there are married couples with both members in the service and assigned to the same installation. Sinop had two to four such situations during my tour. No housing accommodations were set aside for such couples. They could live separately in the barracks, an unusual arrangement, but we did have one such couple. Or they could combine housing allowances and rent something in the nearby community of Sinop. This was referred to as "living on the economy."

All of these variables made life rather complicated. One had to deal with the mission first and foremost and then with whatever the next issue was. Leadership manuals didn't cover these "whatevers." As the astute reader may have picked up, there were two armies on-site—two U.S. Armies. There was the "Old Army" and "Today's Army." And there was a difference between them. And it was a three-letter word that started with an "s" and ended with an "x."

The time frame of this book is 1983-84. The U.S. Army was in the throes of another social experiment, not unlike the one Harry Truman started in 1948 when he desegregated the services. We were now dealing with the integration of women into the Army. The WWII and Korean War model of the Women's Army Corps (WACs) was replaced by one in which approximately 90% of career fields were open to women.

---

2. Turkey - United States Logistics Group Detachment 4

The work we were doing at Sinop was in that 90%, so approximately 30% of Army and Navy personnel assigned to Sinop were female.

The new experiment was effective in 1976. I had a front row seat because, at the time, I was assigned to the Office of the Chief of Staff of the Army and responsible for the weekly General Staff meetings, including taking notes and ensuring that tasks assigned by the Chief of Staff were documented and ultimately completed. Never once during the eighteen months I recorded and followed through on numerous related tasks did I ever think that I myself would be dealing with such cultural change and leadership challenge where the rubber actually met the road at Field Station Sinop.

Another challenge was the Cyprus issue. In 1974 Turkey moved in to occupy the Turkish portion of Cyprus in response to a Cypriot move to take over the entire island. I was on the quick response team in the Pentagon at that time, an unforeseen link to Turkey. U.S. policy argued against Turkish intervention. The Turkish NATO reply to U.S. policy was to put the field station in a lock-down mode. Commanders from 1976 to 1981 had major morale and equipment issues resulting from this technical embargo, and there were still some aftereffects when I took command in July 1983.

So, some thirty years after the fact, I will attempt to illustrate why I regarded that year in Sinop as the best year of a thirty-year career in the United States Army.

Assisting me in the creation of this book were a number of people. The first was my wife, Anne, who willingly encouraged me to take the challenge and risk associated with command and not to play it safe at home. Michael and Rebecca Lustig were leaned upon and responded with valuable insights, as well as data to fill in the gaps of my selective memory. John Knutsen provided feedback on management approaches. Michael Feinberg took my amateur photos and transformed them through his particular magic into the supportive portraits of a beautiful country. And above all else, without Betsy Feinberg's efforts as my editor, none of this would ever have seen the light of day—I resisted and she persisted! Thanks!

By the way, you will eventually understand why the plaque that our aviation department detachment commander, Mike Lustig, gave me as a going-away present was so appropriate. As Mike put it, Sinop stands for **Sex Is Not Our Policy**. Enjoy!

Part I

Laying the Foundation for an Effort at Transformational Leadership

Chapter One

Colonel Hanne? The Call is For You.

I was enjoying one of those false spring days for which central Pennsylvania is famous. It was a bright, clear February day with just a trace of clouds coming in from the west over the top of South Mountain, akin to the movement of Lee's Army of Northern Virginia coming to Gettysburg—quiet and little-noticed, but fearsome in its portent of devastation and struggle, both in Lee's day and, weather-wise, today.

"Colonel Hanne? The call is for you." Janet's announcement interrupted my wondering about future assignment possibilities as my second year of a three-year assignment at Strategic Studies Institute (SSI) was drawing to a close. I hoped that when the time came to discuss options, I would be able to convince MILPERCEN (Military Personnel Center) to leave me here at Carlisle Barracks. Our second son, Brian, had graduated from Carlisle High School the previous spring and was at Shippensburg University, and our third son, Mark, would be graduating this spring. Our youngest son, Eric, would be finishing his sophomore year at the same high school. Our oldest, Matt, was at Fort Polk en route to Berlin later in the summer.

Answering the call, I discovered it was my assignments officer at MILPERCEN, Mark Powe. After a brief but cordial exchange of greetings, Mark dropped a bombshell—would I be interested in taking a command?

I was taken by surprise because not long ago, I had found that I was not on the recently published colonel's command list. Not being selected for an O-6 command had been a disappointment. On the other hand, I wasn't all that surprised. A similar situation had occurred on the O-5 command list. While not a principal O-5 selectee, I had been first on the alternate, but unpublished, list. That's why I ended up with only one year as the G2 (chief intelligence officer) of the 25th Infantry Division in Hawaii. I got there in November 1976, and in March 1977 I was assigned to an O-5 command at Fort Devens, where I would be a battalion commander at the Intelligence School, effective January 1978. That trip from Hawaii to Massachusetts was the third time our family had crossed the country on assignments in four years: Defense Language Institute at Fort Ord, California to the Pentagon, the Pentagon to Hawaii, and now Hawaii to Fort Devens.

I learned from Mark Powe that a similar situation had occurred again. While not on the published list, I was high on the alternate command list and the declination of the command by a principal selectee opened the door for me. Mark seemed to be hesitant in putting forth the question; the reason soon made itself manifest. The command would be in an isolated corner of the world where Anne, my wife, was not allowed to accompany me, a location on the Black Sea coast at a site in Sinop, Turkey.

Of the many command possibilities that existed, Sinop was considered the least desirable, if for no other reason than it was low key and not very visible. That meant it was not regarded by aspiring colonels as a stepping stone to selection as a general. In addition, whereas all other commands were 18 months in length, Sinop was only 12 months. Selection boards were allegedly of the opinion that most of us could carry off a 12-month assignment standing on our heads, but it took 18 months for all the "chickens to come home to roost." In other words, it took 18 months for mistakes in leadership to catch up to you; plus there would be at least two effectiveness reports on actions taken: the annual report and the end-of-assignment report. On the other hand, there would be only a single end-of-tour report for Sinop, reducing the opportunity to show off the necessary skills for the selection committee to consider you for promotion to General.

I knew all this. Still, it was a command and an opportunity to serve in a new and challenging position. I would have the opportunity to exercise the leadership theories I had developed over the years, and of necessity, I would have to live with the results. There was only one potential obstacle to saying yes—Anne. That obstacle could be overcome if it would be possible to return to Carlisle Barracks and SSI after the tour in Turkey. Reassignment to Carlisle would mean that the family would remain in the Carlisle area. Eric would graduate from Carlisle High School, where he was doing quite well, and we would be spared another move. Carlisle Barracks had been move number 18!

On a personal note, I always believed that my name was against me. Senior officers who knew me by name fell into using Hannah as the pronunciation, eventually forgetting that the pronunciation might be Hannah, but the spelling was Hanne. And those officers who knew me only from the file of effectiveness reports would call me "Haney," and the two—Hannah and Haney—never connected. I was passed over seven times for brigadier general. But such is life!

Mark understood that if Major General Lawrence, who was the Commandant at the War College, the senior command at Carlisle, and my current boss's boss, agreed to my return to SSI or the War College as a potential instructor, there would be no problem in accepting the command. I told Mark I would call him back after lunch.

My next move was a visit to the Commandant's office to get the required approval. There was no problem, as I had been a successful researcher and writer on matters of interest and concern to the General Staff. So with that in hand— literally, since it was a personal note from Major General Lawrence agreeing to my return—it was off to lunch and breaking the news to Anne that we were facing one more extended separation if she accepted my taking the assignment.

Anne turned out to be as elated as I was over the opportunity for command. She had been disappointed when the primary list had been published and I was not on it. Mark's call had suddenly opened new vistas for all of us. Yes, this was another separation, but this one did not entail the war-time risks that two previous year-long tours had brought. We agreed that Eric's driver training would be postponed a year, so that she would not go through what I had with his three older brothers—one learning in Massachusetts, one learning in Newport, and one finishing up here in Pennsylvania.

Then we learned that there was a fly in the ointment. We were in government quarters because I was a member of the staff and faculty at the War College. We assumed—spell assume *very* slowly—that we would be allowed to stay in the same quarters, since Anne was specifically denied joint travel. Wrong! In retrospect it made sense. Carlisle Barracks had limited housing. With each graduating class having its fair share of one-year "short tours," allowing stay-behind families to remain in quarters would really bollix up housing. We eventually found a place to rent in Carlisle and made the move—Number 19—in late May.

Having made the call to Mark after consulting with Anne, I was brought back to reality as Mark spelled out what came next. One did not just walk into a command position without preparation, and my case was no different. Between February, when the sequence started, and July, when I took command, there were a number of mandatory meetings and courses in which I had to participate.

The primary Defense Department agency involved with Sinop was not the Army, so several trips down U.S. 15 and Maryland Route 32 from Carlisle Barracks to Laurel were necessary. After that, there was a two-week session at Fort Huachuca, Arizona, where we ("we" being other Military Intelligence command selectees, lieutenant colonels and colonels) got exposed to the most recent changes underway within our Military Intelligence (MI) branch.

The visit to Fort Huachuca gave me an opportunity to become familiar with the Sonoran Desert and the activities available in the area, not expecting that in less than ten years, Anne and I would be retired and living in Cochise County, to say nothing of our move in 2013 to Green Valley…karma is something else!

And so was the Mule Mountain Marathon (MMM): 26.2 miles from Bisbee, Arizona to Fort Huachuca. There were five of us—the Iron Colonels—who organized a relay team in which each member would run a five-mile leg, except number five, who would do the six-plus miles to the finish. Here was another incidence of karma; I ran in the 1983 MMM and ran it again as a member of the Huachuca Hash House Harriers from 1991 through 1995. The last MMM was in 1996 when Anne and I were in Croatia, where I ran with the Harriers in Zagreb. Traffic had become a real hazard and a source of complaints, so the MMM ceased in the name of safety.

All coordination visits were interesting and informative, but the real highlights were the Judge Advocate School briefing and the U.S. Army Intelligence and Security Command (INSCOM) sessions.

The Judge Advocate School session was short, but especially meaningful. The Judge Advocate General is the Army's lawyer. Because of the complexity involved with the concept and conduct of command in itself—the legality and responsibility involved—two points stuck with me. The first was more practical than legal: One of the many briefers noted that before one made a policy change or took action, it was important to consider what that action would look like on the front page of *The New York Times.* In other words, stand back and attempt to look at the situation from a non-emotional point of view. As my father once said, paraphrasing the classics, "Whom the gods would destroy, they first make angry."

The second concept that made a lasting impression on me was the reminder that any actions or statements one makes as a commander are subject to providing sworn testimony, especially in courts-martial cases. It is critical to understand that any decisions one makes as a commander carry with them the possibility of testifying under oath. The old expression "what a tangled web we weave when we first start to deceive" comes to mind. This is emphasized by the almost daily news articles on how cover-ups usually result in worse situations and end results than if there had been openness and clarity from the very beginning.

INSCOM was an entirely different experience. The organizational structure, command relationships, and operating procedures, dull in their own rights, became even duller and less memorable when compared to the team-building exercises we went through. The INSCOM commander at the time was Major General Albert Stubblebine, rumored to be Lee Marvin's doppelganger. Stubblebine was into the esoteric. In fact, Jim Channon, the lead character around whom the movie *The Men Who Stare at Goats* was structured, was a principal advisor to Stubblebine on the team building program IBEX. (See Jim Channon's MI cartoon on the next page.)

IBEX was derived from "INSCOM Beyond Excellence" and was based on the logic that with pending financial and personnel drawdowns based on the then-current recession (1983), INSCOM would have to do more with less, so innovative approaches and techniques must be considered. The concepts we were exposed to ranged from neurolinguistic programming techniques used during interrogations to concentration of mental energy in "spoon-bending" exercises. This was a time when mental telepathy was in vogue, and there were almost weekly demonstrations of the projection of mental images or the powers of mental telepathy. We were exposed to state-of-the-art ideas on what made a team function and how to develop teams and build teamwork so that there was a synergistic effect derived from the cooperative and collaborative attitudes. There will be more discussion on these points as we go through these reminiscences from a thirty-year perspective.

Keep two concepts in mind as we move along. The first concept is to decentralize power and the authority to make decisions down to the appropriate level. The key term is "to power down." (This does not mean decentralizing the responsibility. That is always the commander's—he or she can delegate *authority*, but not *responsibility*.)

The second principle of IBEX leadership is to give key players the freedom to fail. If initiative on the part of subordinates is desired, then there will be an element of risk involved. The quickest way to kill initiative and stifle subordinates is to castigate and penalize them for honest mistakes. (Ever since, the most oft-heard saying in our family has been, "Well, what did we learn from that?" That idea lingers on today in our son Eric's Middle East History courses at Florida Atlantic.) Note that "freedom to fail" does not include "freedom to be incompetent." There is a critical difference between initiative and incompetence. Experience beats book-learning every time!

ASRAC the Magnificent
HI SIDE
ARMY SECURITY AGENCY
PIGMY TRIBAL
GRIN & BEAR IT
MASS
POWER & LIGHT
CO
MI QUALIFIED DAMN RIGHT
GI
G-2 DRESSIN
BALLET SHOES
SOFT SHOES
GUIDE TO NEW HOMES
MILITARY INTELLIGENCE

SHOW
WHO'S WHO
AGGRESSOR LAND
3 SHOTS FOR 10¢
DUNK THE DEUCE
AIR MOB
LITTLE BIG HORN
SHORT ROUND
STP
CHANNON
THE GREATEST SHOW ON EARTH

The time went quickly from Mark Powe's call in February to my flight at the end of June to Frankfurt en route to Istanbul. Our son Matt's assignment to Berlin was confirmed, and he was able to spend some time with us in Carlisle before leaving. Brian was engaged in Army Reserve training in between semesters at

The Hanne family - L to R - Matt, Anne, Mark, Bill, Eric, Brian

L to R - Matt, Bill, Brian

Shippensburg. Mark had graduated from Carlisle High and had enlisted in the Army as a medical technician. Our youngest son, Eric, was left wondering at the change: where there had been a family of six always coming and going, there was now a cell of two keeping watch over home base while the rest of us were "up, up and away!"

Chapter Two

Initial Impressions—Both Directions

The flight to Turkey was uneventful. I flew out of Dulles, with Anne and Eric in their usual roles of seeing me off on another adventure while they kept the home fires burning. This was well before the days of e-mail, so we knew that all personal contact would be on a two-week turnaround basis. Fortunately, the Sinop site had a flight detachment of two Beechcraft Queen Airs (U.S. Army C-12s) that flew to and from Istanbul daily, picking up and dropping off mail and passengers: troops arriving or departing the field station. Because we controlled flights in and out of the area, we were never surprised by unannounced visitors from our next higher headquarters.

Beechcraft Queen Air (U.S. Army C-12)

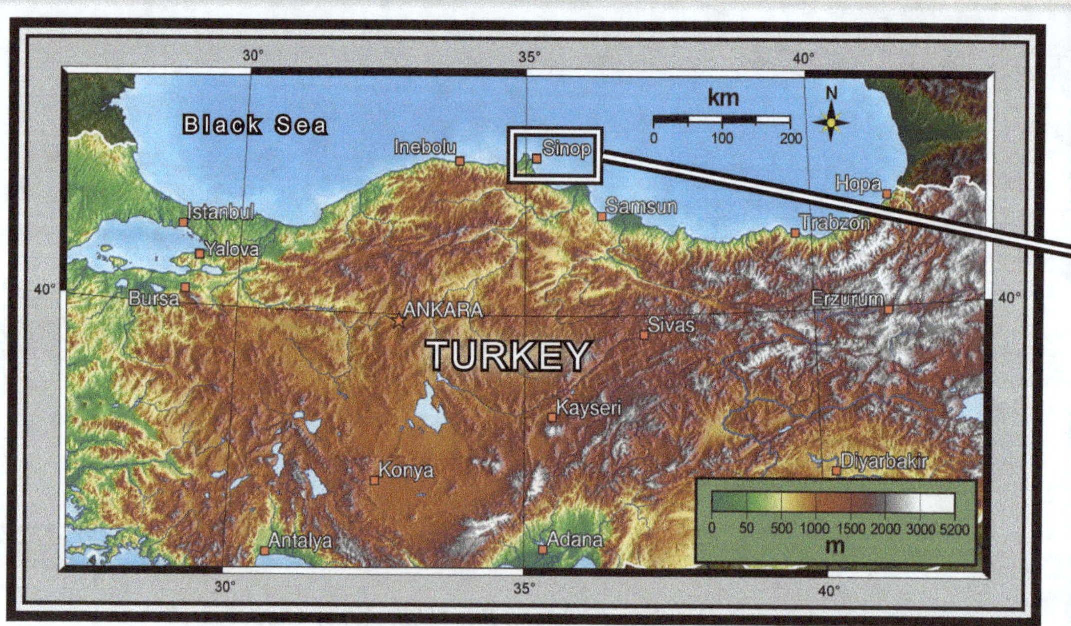
km
0   100   200
N
Black Sea
Inebolu
Sinop
Hopa
Istanbul
Samsun
Yalova
Trabzon
Bursa
ANKARA
Erzurum
Sivas
TURKEY
Kayseri
Konya
Diyarbakir
Antalya
Adana
30°
35°
40°
40°
40°
0   50   500 1000 1500 2000 3000 5200
m

he view west from Diogenes Station toward Sinop

*The view east from Sinop
toward Diogenes Station*

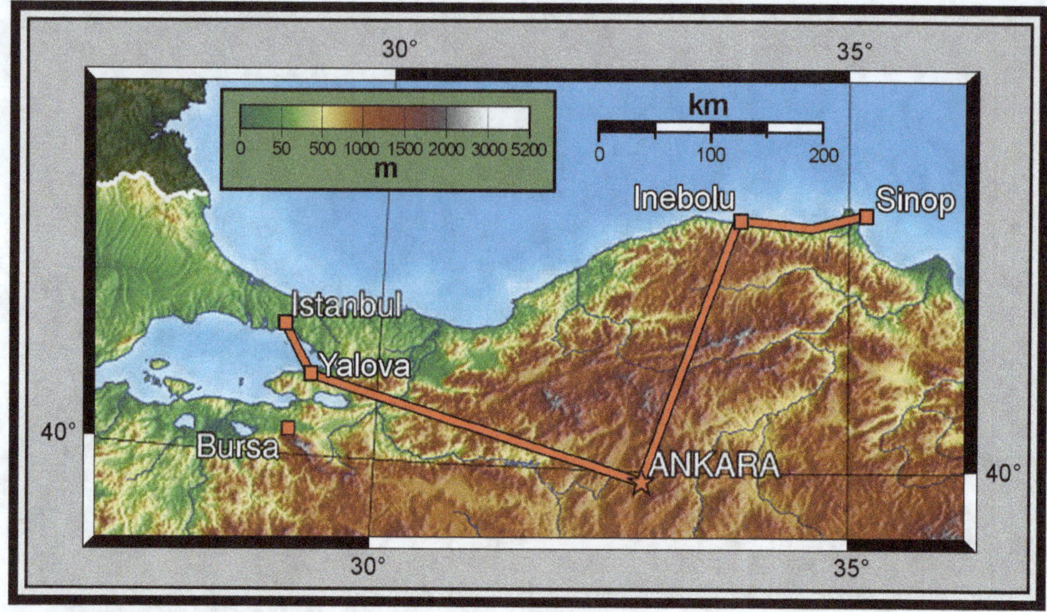

Telephone communications were also available, but at a cost. (There was no charge for official calls; after all, we were a NATO communications site.) The flight time was two hours over a prescribed route—no freewheeling in this neck of the woods! The flight path was from Istanbul to Yalova, Cebuk (Ankara), Inebolu and Sinop.

Even the Air Force C-130s that kept us resupplied with everything from food to technical devices flew the same route. And ours was not an easy take-off and landing strip. There had been several no-survivor C-12 and C-130 crashes into the mountainside overlooking Sinop.

The physical and human geography of Sinop is key to understanding its strategic position and its importance throughout history. As the map on page 16 illustrates, Sinop is essentially

Lockheed C-130 Hercules

the remains of a mountain that juts out into the Black Sea. Deep-sea explorer Robert Ballard and others have explored its depths, and conventional wisdom holds that in about 7,000 to 8,000 BCE, the meltwater from the snows and glaciers flowing into the Mediterranean overwhelmed the land in the vicinity of the Dardanelles and the Straits of Bosporus. This started a major flow into the basin, which at the time had a small lake in its lower reaches. The flow was of such speed and intensity—estimated by some to be enough to cause a rise in sea level of a foot a day—that the bottommost oxygen-free layer of water—the original lake—has stopped any decomposition of villages and structures found at the bottom of the sea. Because all historical cultures surrounding the Black Sea make note of a great flood, many researchers believe that this inundation caused by the overflow from the Mediterranean is the starting point of the Biblical story of Noah and the ark.

The site itself is one-of-a-kind, but if you throw in the history of the region, the entire area takes on a unique aura. The Hittites were the first to use the port of Sinop, but based on the bay's layout, you can be sure that any people using the Black Sea as a means of travel would find safety in the bay. The prevailing winds are from the west, and the isthmus on which the city lies provides a safe harbor for any type of vessel. While the Hittites may be the first recorded users, it is logical to expect that the sequent occupance[1] of the area predates the Hittites.

Around the 7th Century BCE, the Greeks built up the area into a major Black Sea port and colony. The area was taken by the Persians in the 4th Century BCE and then by the Romans in 70 BCE. Julius Caesar established a colony at Sinop in 47 BCE. When the Roman Empire split in 395, Sinop became part of the Eastern (Byzantine) Roman Empire.

The next notable invasion came with the attacks by the Arabs in 858, most likely bringing Islam with them. In 1081 the Seljuk Turks captured the city, only to lose it to the Byzantines until the fall of Constantinople in 1204, when Sinop was captured by the Empire of Trebizond. The Empire ruled Sinop until the Seljuk Turks recaptured the port. The Seljuks retained control off and on until 1265, when the Turks established control and remained so until the present time. The walls of the city today, including the gateways, tell of a vibrant and contentious history.

The walls of Sinop

---

1. Sequent occupance is a term in human geography expressing the idea that successive societies leave their own imprint on an area, each contributing to the cumulative cultural landscape.

One aspect of Turkey that fascinated me as a geographer was the cultural change brought about by Mustafa Kemal Ataturk in the 1920s. One of the basic principles of human (cultural) geography is that changes to a nation's culture are brought about from the bottom up, not from the top down.

Ataturk turned this notion upside down. Under his direction, Turkey transitioned from an Islamic nation to a secular state. The written language changed from Arabic script to the Roman font used in English. The fez was replaced with a hat with a brim. Among the major cultural changes was the advancement of women in all professions. Turkey had been called the "Sick Man of Europe," a term first used in the mid-nineteenth century to describe the Ottoman empire. Ataturk changed that. He transformed his country.

The Shah of Iran tried to make the same changes to Iran's culture, and we know the results of that effort. In some ways, as I write this in the spring of 2016, the vote may still be out on Ataturk's changes, given Turkey's recent moves away from a secular state. In 1983 the changes of the 1920s appeared to be lasting, demonstrating that the top-down approach can work. But today? We have to wait and see.

Ataturk's mausoleum in Ankara is impressive, as are the statues lining the walkway to the tomb. On one side are statues of three men: a soldier, a peasant, and a professional. On the other side are statues of three women: a teacher, a peasant, and a professional. These represent the three major activities upon which the state and the welfare of the people existed: the farmer, the educator, and the soldier. It is all very impressive. But back to Sinop...

Ataturk's mausoleum in Ankara

The statues lining the walkway to Ataturk's mausoleum.

Three men: a soldier, a peasant, and a professional.

Three women: a teacher, a peasant, and a professional.

The field station site is some 700 feet above sea level and the actual city of Sinop. It sits on a large plateau where the wind never seems to stop blowing. Winter storms are wicked—so wicked that it was necessary to place ropes for safety among the many buildings of the field station. The troops and workers used these ropes to provide support and guidance in the often blizzard-like or white-out conditions that dominated the winter weather. The storms would blow in from the west and the north, come up against the 6,000-foot mountain range to the south and dump massive amounts of snow in the winter and rain in the other seasons.

Field Station Sinop - Antenna Array

Field Station Sinop - Support Buildings

The mountain ranges were steep and heavily forested due to the ample amounts of orographic precipitation.[2] There were only a limited number of passes through this range and their winding highways leveled out on the coastal plain at Samsun to the east and Karabuk far to the west. The coastal highway ran from Istanbul in the west to Trabzon on the east. There the highway shifted to mountain passes before reaching the Soviet (now Georgian) border.

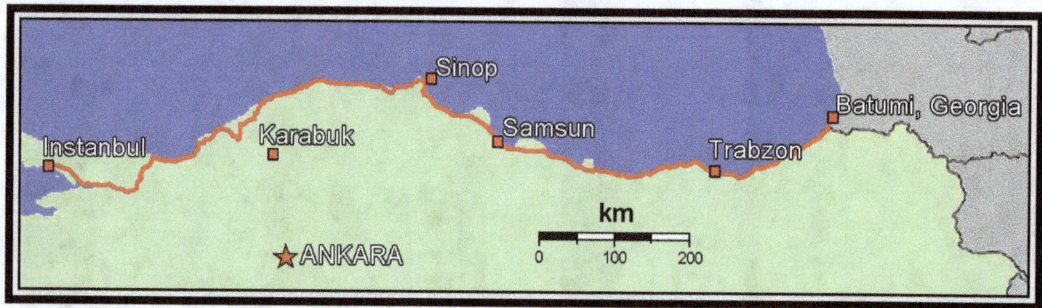

This coastal area was literally a separate Turkey where Greek and Roman remains competed with Seljuk and Ottoman Turkish heritages.

Costal area walls and artifacts

---

2. Orographic precipitation.   Rain, snow, or other precipitation produced when moist air is lifted as it moves over a mountain range. As the air rises and cools, orographic clouds form and serve as the source of the precipitation, most of which falls upwind of the mountain ridge.

Sinop proper is a major port on the northern coast of Turkey, oft-visited by tourists and cynics—this was Diogenes' home town (more on the role of the museum in this story later). The coastal area is fairly broad and dominated by agriculture, fishing, and boat-building (mainly traditional Black Sea fishing boats).

During my tour, the "White Boat," owned and operated by a Turkish travel company, would sail once a week from Istanbul to Trabzon and back. This proved

Port of Sinop

Model of a Black Sea fishing boat

to be an excellent way for the troops to take advantage of a short leave or break in routine—to leave on the White Boat on Wednesday and return on Saturday. There was a crystal glassware factory not too far from our site, as well as a fish processing plant. The city had several hotels, the newer ones being visited by troops on their time off.

Having one's own surgeon and dentist as part of the U.S. Army team helped in keeping track of command indicators like STDs and intestinal ailments. Neither issue—the extent of STDs or stomach ailments—was considered to be a problem by our surgeon. Common sense would have it that both existed to some extent, but not to the point where "something had to be done!" We also had two chaplains, one Roman Catholic and one protestant (Episcopalian) during my tour. Enough about routine military issues. Back to my flight from Washington to Sinop.

The flight to Rhine-Main was by a U.S. flag carrier, but it was Lufthansa from Frankfurt to Istanbul and Army Aviation from Istanbul to Sinop. My memory of the flights is minimal, which means that they were on time, smooth, and no luggage was lost. One memory stands out, however. This was the first encounter I had with fully-armed, locked-and-loaded patrols inside air terminals. This was Europe in 1983. We Americans were still unaffected by terrorism, but Europe, especially Germany, had experiences which we had not. Our security patrols inside Dulles were mainly overweight, overage, and overtime, with personnel more interested in watching the women going up escalators than in almost anything else. This was in stark contrast to the uniformed police patrols, often complete with canine companions, that seemed to be everywhere in the terminal at Rhine-Main.

I was on the lookout for Americans on the Frankfurt-to-Istanbul flight, as there might be one or two others headed for Sinop. (We wore civvies instead of uniforms for personal security. The rule was "Don't draw attention to yourself!) I knew that Major General Stubblebine would meet with me when we got to Istanbul, but there might be others onboard who were headed for the field station.

MILPERCEN had asked me back in the beginning if there were anyone I wanted to go with me. On a one-year tour, there was constant turnover. Eventually everyone who was in place when I took command would leave before me, so I had the chance to choose their replacement. I declined the offer. I was willing to take whomever MILPERCEN sent me. My deputy, Lieutenant Colonel (LTC) Norm Chung, had already been onboard for three months. The third member of the command team, Command Sergeant Major (CSM) Russ Welker, had also been there for a couple of months. The command team would be complete with my arrival. I saw no one on the flight who could be positively identified as Sinop-bound.

I was met by our liaison personnel at Istanbul and spirited away to accommo-dations at the seafront Kalyon Hotel *(Kalyon Oteli)*, where I met MG Stubblebine for dinner. He would be present at the change of command between myself as the in-coming commander and Wayne Stone, the departing commander. The menu that first night is a first-rate memory—open-grilled fresh-caught flounder with rice pilaf and a salad of tomatoes and lettuce—a capital meal and a great introduction to yet another cuisine.

MG Stubblebine and I met the flight-detachment commander at the Istanbul terminal. There may have been one or two others on that flight, but if so, I do not remember them. This took place in 1983, and it was 2016 when I finally decided to share, for what it is worth, my perspective on principles of leadership as I experienced them under unique conditions. In any case, the flight into Sinop went smoothly.

Wayne Stone, the out-going commander, met us at the airstrip per proper protocol and escorted us up to the Field Station Headquarters. There we met Colonel Senal, the Turkish commander. Having traveled from the States in civilian clothes, I was taken to what would be my home for the next twelve months, so that I could change into a uniform. At the time that would have been Army Greens: green blouse and trousers, shirt and black tie, and low quarters. Remember this detail. As a Russian playwright is alleged to have said, "If you have a rifle over the fireplace in the first act, it had best be used by the third." After changing, I was to get a view of the remainder of the site and meet some of the key personnel. Among those personnel were LTC Norm Chung, my deputy commander, and CSM Russ Welker.

I do not remember the exact timing of events, but some time that afternoon or the next day, we had a formal change-of-command ceremony in which Wayne passed the American flag to MG Stubblebine, who then passed it to me. Although a very simple ceremony, it was highly emotional. Following the review of troops, MG Stubblebine made a few remarks. He was followed by Wayne. Then it was my turn to pass along the vague, glittering generalities normally associated with such ceremonies. Having raised four sons, three of whom were in the service at the time of my takeover, I knew from experience the probable attention span and general interests of the troops standing at parade rest during these remarks. Consequently, mine were short, sweet, and to the point—with the key point being "stay tuned, there is more to follow."

There was no doubt whatsoever that while I was assuming command I was being carefully scrutinized by everyone associated with the field station—everyone from the Turkish laborer who worked on our water lines (an item of interest) to the Turkish cooks who provided us with four meals a day,[3] to Norm and Russ, who wondered what to expect from this newbie who nobody on-site had ever met or knew anything about, but who was going to have a tremendous influence on their lives and existence as long as they—and I—were associated with Diogenes Station.

---

3. We ran a 24-hour operation so that the graveyard shift would have access to a late-evening meal before going to work.

Chapter Three
Keeping Your Eyes Open and Your Mouth Shut

Wayne left immediately after the change-of-command ceremony. The king is dead; long live the king!

The Army's policy of not leaving the past commander on-site is practical and smart. There are, and always will be, those who are drawn to power, either to their own or to someone who possesses power that they enjoy being associated with. This personality type can have extreme potential for harm, so with the departure of the old commander and the arrival of the new, all former relationships are dissolved and new ones started.

My first stop after the ceremony was to meet with Stubblebine about some last-minute guidance-and-command information. He did not have much to say, other than to express worry that I might be too easy on the troops and a pushover *vis-á-vis* discipline. Where this concern originated is unknown. I can only guess it was based on concerns of contemporaries at INSCOM HQ who did not know me, but recognized me as a threat. Really? A threat? How? I was not from the major headquarters staff clique, but from the tactical side of the house and SSI research centers where military intelligence officers were few and far between. I was a 25th Infantry Division G2, a chief intelligence officer in a fatigue uniform and muddy boots rather than an officer in Army greens and spit-shined low quarters. However, one has a tendency to fear what one doesn't know. I assured the Major General that there was nothing to be worried about, but even today I remain certain, based on his body language, that he was not convinced. But what the heck! He'd soon be wheels up and out of my sector.

As he and Wayne took off in one of the C-12s, we—Norm, Russ and myself—headed back up The Hill[1] to a session in my office in which all the key issues were laid out for discussion. The immediate question concerned when I would brief the troops and lay out command policies and goals. Everyone who had briefed me about the site had painted a positive and rosy picture, but now the gloves were off. I had to learn the nitty-gritty of running a joint headquarters in a foreign territory that was a two-hour flight from the next highest headquarters, and I was responsible for over 500 personnel, one-third of them female. So far, there was nothing I hadn't expected — until Norm noted that there were indications that up to one-third of our officers—married and unmarried, warrant and commissioned—were in bed with enlisted personnel (At the time there was only one female officer on the team. That number would grow threefold in the coming months).

Now we had a problem, plain and simple, and it was called consensual intercourse within the supervisory chain of command.

The social experiment that had started in 1976 with the integration of women into all but combat specialties[2] found the armed services unprepared to handle "what comes naturally" when you mix men and women on the job, especially 18-to-24-year-old men and women—single or otherwise. Some of those in the discussions during the pre-deployment workups offered the approach of simply stating, "No sex. Period." My response was, "You've got to be kidding if you think that is going to work!"

Anyway, the problem—if we had one—was now mine. Next?

Micromanagement. Apparently, my predecessor was a Type A personality who insisted on knowing what was going on at all times—to the point where the early cell phone was glued to his hand and never left it. (In 1983, cell phones weighed so much they resembled a brick with an antenna.) The most-used number was the U.S. military police location just inside the compound. Evidently, the MPs were always on the run looking here or there for this or that or checking out bogus reports. Compared to the sex issue, this was a no-brainer. Next?

Alcohol. You could enlist at 18 and become an alcoholic by age 19. Many troops had no idea where Turkey was or what existed outside the gates, and too many—and even one was one too many—would crawl inside a Jim Beam bottle on their time off. This was a problem and was most likely a factor in the reputed death and suicide rate associated with an assignment to Sinop. Although that was never proven, it was

---

1. The site had a capitalization quality to it that was based on its physical dominance of the immediate area and on its position as a NATO site. We often called it "Hippodrome Hill."

2. There were even some exceptions in that area, especially in combat electronic warfare and intelligence positions, such as the collection-and-jamming platoon leader, who was frequently positioned in front of combat infantry positions.

always part of the legends associated with the field station. You might not be at the end of the world at Sinop, but you could see it from there.

This was one of those "gotcha" problems. If I, as commander, said "We have a problem," without anything solid to back it up, then we did have a problem. That's just the way things worked. I needed a solution to a problem that—as far as the troops were concerned—didn't exist. I must have slept through that subject in one of the many management courses Uncle Sam sent me through before turning me loose in Turkey. This one was going to require a little thought. And I had to remind myself that I gave up thinking as "conduct unbecoming an officer," when I was commissioned. Next?

Keeping the troops busy. Routine hours for the Army communications personnel were thirty days 0800 to 1600 (24 hour clock) for the day shift; two days off, then thirty days 1600 to 2400 for the night shift; end the cycle with two days off, before thirty days 2400 to 0800 for the graveyard shift, and then a week off before resuming the cycle. The Navy used a two-two-and-eighty routine: two 12-hour days followed by two 12-hour night shifts, followed by 80 hours off. There were some suggestions made at the first command briefing to get the Navy to follow our approach. I killed that conversation right there. The U.S. Navy had its own approach to things, and I saw no logic in butting my head against a wall, a procedure that was working and caused no issues for us. End of conversation. But there *was* a problem related to boredom, and the alcohol issue was an indicator of a relationship, so some thought had to be given to this matter. Next?

Norm and Russ felt we had hit the high points that needed immediate attention. Both of them suggested an all-hands briefing in the theater in which I would share with the troops my command philosophy and goals for the organization. I agreed and asked them to set things up for a theater briefing in two days. Meanwhile, the three of us would work through these issues of sex, micro-management, alcohol, and troop activity, keeping in mind that the mission came first.

The end of the work day had come, and many of the officers had gathered in the DOOM—Diogenes Officers Open Mess, i.e., the officers' club. (There was an enlisted club and a non-commissioned officers club on-site, as well as the DOOM.) I am sure they knew where I had been and to whom I had been talking, and now they were looking for their turn to find out what I was like in an informal setting. I started with a tonic and lime and stayed with it, not only that night, but for several weeks thereafter. Inadvertently, I caused Norm all sorts of problems because I'd listen to what an individual had to say, and I would nod in agreement, meaning that I had heard what he or she said. Norm explained later that evening that my nod was interpreted as a go-ahead for whatever suggestion might have been made.

Lesson Number One: Don't agree to anything in the club; just listen and say, "I'll get back to you." about that suggestion, idea, or whatever. Lesson Number Two: Have a deputy who is professional enough to read what is going on, what precedent I was unwittingly setting, and tell me about it in a professional way. Norm saved our hide more than once by being the professional that he is.

After a couple of tonics and lime, the stress of the day caught up with me, and I was off to what would be my home for the next 365 days. Home was a suite of two rooms: a sitting room for formal occasions and a bedroom with wall lockers for uniforms and other clothes. I fell asleep wondering what I was going to say to the troops in a couple of days. If tonight's exposure to the officers in the DOOM was an indicator, there wasn't much room to maneuver. There were too many people watching; it was like a small town where everybody knows everyone else's business. And it was typically Army: everyone was sure they could do a better job than the boss was doing.

Chapter Four
Getting Established but Without Getting Into a Routine

The next morning, after a shower and shave, I got into my BDUs (battle drill uniform: the old camouflage fatigues) instead of the greens I had worn the day before. I stopped by the mess hall and walked through the kitchen, where I introduced myself to each cook, shook hands, and asked about their job—basically, making small talk. I had learned a long time ago to focus on the 3Ms: mail, money, and mess. Getting paid on time and in the correct amount and having a good mess-hall experience was essential in keeping the troops' morale up. As I went through the chow line, I spotted several officers sitting together, so I joined them. We picked up where we had left off the night before, exchanging details about our careers.

I learned later that the news of my wearing BDUs, rather than the greens that the other officers wore, and my going through the kitchen before going through the chow line, made it through the compound and field site faster than wildfire. The general consensus was that there was going to be a new approach taken by the command. What I was slow in doing was putting myself in their position. Every time I had been involved with a new commander or supervisor on the scene I had been like these officers, sizing up the new boss and learning how to approach him. It's just human nature to do so.

I attempted not to follow a set pattern, believing in the approach I learned from my first troop commander, Art Strange. His philosophy was to be where your troops were and to avoid following a set pattern. That way you would keep them on their toes. Later I learned that this technique was known as MBWA, "management by walking around," a concept I will elaborate on further.

As a matter of routine, I would first check in with my administrative assistant, Specialist Five April Hunt. She kept my calendar and intercepted phone calls, visitors, and potential problems. She would take them to Norm or Russ first because they had been on-site and were aware of old and on-going issues. They would bring me up to speed in due time when it was necessary or when the issue was significant. Norm and Russ bought into the "power down" and "freedom to fail" philosophies from the start and relieved me of many minor issues which, if brought to me first, would only have been bounced back to them to solve.

Depending on what was scheduled on my calendar, Command Sergeant Major Russ Welker and I would walk through the barracks at least once a week. (There were always off-shift troops sleeping, regardless of when I walked through.) We would hit the admin offices daily and meet with the Turkish commander once or twice a week. Every two weeks or so, we would go downtown and meet with the governor, the mayor, and/or the authorities responsible for criminal and related issues. We would kibitz over a cup or two of Turkish coffee. I took mine straight; they had difficulty believing that I would not use sugar. We were respectful of each other's position in the community and the special relationship that existed between them and the NATO site.

The value of the relationship established over innumerable cups of coffee paid off several months into my tour. The movie <u>Midnight Express</u> was still fresh in everyone's mind. It was a rather graphic film on what one could expect if arrested for drug use in Turkey. One morning at coffee, the chief of police casually mentioned that under the provisions of the DECA/SOFA, he would be visiting our compound in a couple of days to arrest Specialist X on a charge of selling drugs. He produced photos taken of the exchange of drugs for cash. I thanked him for the advance notice of his visit and casually continued with the "coffee call" in the usual manner. Returning to the compound, I told CSM Welker what had happened. He got the young troop investigated. When we told Specialist X what was coming next, he readily confirmed that he had been selling marijuana on the side. That afternoon we shipped him off to Augsburg with the necessary paperwork for proper legal proceedings.

The police arrived as scheduled, and the chief expressed regret that Specialist X was not present. But he was also glad that we were handling this within our channels, and that it was not necessary to make it an international incident. I am convinced that those mornings of coffee-drinking and casual conversation paid off in the deliberate heads-up by the police chief. He did not want the publicity associated with the arrest of an American, but he had the law to uphold, as did I.

On the lighter side, the governor asked one morning what I had done wrong. I didn't understand his question, nor did my interpreter, Octay. The governor then explained that all the political authorities, legal officers, and even the military commanders assigned to Sinop were there in a sort of in-country exile. The Turks thought that maybe the same thing applied to the U.S. officers. (Colonel Senal was not with me that morning.)

In a couple of weeks I got to the point where I knew where I was and what the troops were doing. There was no interference on my part, but it did take the troopers, especially the non-commissioned officers (NCOs), some time to get accustomed to my method of MBWA.

CSM Welker explained the rationale behind the NCOs' concerns. They were of the opinion that I was listening to the trooper more than I was to them. Russ described a typical scenario: I would appear on the scene right after Soldier A had been eaten alive by his supervisor for some dumb stunt, and the first person I would start talking to was Soldier A. He would pour out his tale of woe, painting the NCO in hues akin to Attila the Hun. I would take it all in from the soldier's perspective, leaving the NCO in an unfavorable light. It took a couple of months, but both the NCOs and the soldiers learned that before any action was taken, both sides—or as many sides as there were to the issue—would be looked into, assuming that it was worth the time and effort to hear them all, as dictated by common sense.

My MBWA approach saw me visiting the communications operations site virtually any time of day or night, ensuring that both the supervisors and the troops involved knew that their efforts were appreciated and recognized. While I don't remember ever taking any punitive action based on what I learned, I did gain insight into their world, their concerns and challenges. Most importantly, it was from them that I first got wind of the never-ending rumors that sweep through any organization. I learned that the most correct and immediate action was to spike any rumor by either providing the truth or denying it, eyeball to eyeball. Rumors, if allowed to go unchecked, will destroy morale faster than anything else the commander can do.

My day quickly fell into a typical routine: a run sometime during the day,[1] a quick what's-new session with Russ and Norm, usually on separate topics (easier than carrying that "brick" around), and being seen by the soldiers and sailors as they went about their duties. There was no hiding behind my desk, except

---

1. I belonged to the 2000-mile club by the time I left the command.

for the formal historic photo of the commander, that was hung next to Wayne's picture and just ahead of the space for the photo of my replacement, Ted Fichtl.

The formal photo of the commander

I always paid a visit to the mess hall, whether I had lunch or not. And I always had supper there with whoever was around. It made no difference who it was because we would always find something to talk about, be it the weather or the poor choice of movies we received through the system. I would hit the mail room after supper. More than once I encountered a troop who was despondent over his "air mail," i.e., an empty mailbox. Anne and I had plenty to write about, so there was usually a letter waiting for me, much to my joy.

I capped off the evening with a quick visit to the DOOM, continuing with the tonic and lime routine for the first several weeks. The night that I casually ordered a 7 and 7, you would have thought that the officers around the bar had seen the end of the world. That night several of them admitted that they had thought I was a teetotaler. Now they understood that I was setting a standard for alcoholic intake. It was common knowledge that I quietly tracked sales of alcohol. That in itself had a tempering effect, and sales slowly went down over the year. It only goes to show you that while culture usually changes from the bottom up, effective leadership starts at the top and goes down.

So far, so good, but now it was the night before my briefing to the assembled commands—U.S. Army and U.S. Navy—and my focus was turning to what I was going to say, especially about the number one issue, sexual behavior.

Chapter Five

Or is it Sex?

I felt it was important to communicate the major aspects of my command philosophy in a positive way. I knew from personal experience, as well as from war stories told by our sons based on what they saw, that a list of "do's and don'ts" would put the troops to sleep, and it would be a wasted afternoon. So I capitalized on some of the ideas I had heard in our series of presentations and classes at the Army War College and the Naval War College. Rather than put stress on the obvious, I would present my "magic sandbox theory of leadership," followed by a number of key points that I would be emphasizing during my tour.

A major issue was cohabitation. Call it what you will, but being the father of four, I could not logically or ethically say I was against sexual relations—but I *could* say I was against fraternization.

And what was fraternization? Fraternization was playing favorites: showing, or even giving the appearance of, favoritism to subordinates. If you were in a command position in which you would make decisions on promotions, awards, and working conditions, or if, as a supervisor at any level, you had control of planning, organizing, supervising, deciding, coordinating, budgeting and training, then showing favoritism was the quickest way to destroy the cohesiveness of the organization. In the armed services, the primary objective is the mission. Anything that takes away from accomplishing the mission is *verboten*! But to stand up in front of the assembled troops and say "no more sex" would mark you as unfit for command, based on your absence of common sense. On the other hand, by putting the mission first and making the absence of fraternization the standard, I might stand a chance of some degree of success in building and maintaining unit cohesiveness and in attaining our mission goals.

All these ideas were based on the assumption that the sex was consensual. It assumed that both parties agreed to bedding down together. But what if Party A did not want to bed down with her supervisor? What if the word came out that unless she did, she probably would not be put forward for promotion, or her hours would change to the permanent graveyard shift? That puts an entirely different complexion on the problem.

I went for a run before the briefing. It was my way of clearing the brain, as well as an opportunity to mentally rehearse. I used 4x6 note cards with keywords as the prompt. As I had already used most of the material during leadership sessions at the Army War College, everything but the sex/fraternization issue was old hat.

I waited backstage until it was time for Norm to introduce me. Once the troops were seated, I launched into my leadership approach. While not word-for-word what I said that afternoon, the following is based on my rehearsal notes, most of which are still in my possession. That presentation follows:

"Good afternoon. I am Colonel Hanne, your new field station commander, and I would like to take a few minutes to share with you what I regard as my approach to being a commander.

First of all, let's get the name straight. The last name may end in an "e," but it is pronounced "a." So, it isn't Colonel Hain or Colonel Han. It is "Hanna, but spelled with an "e." You are confused? Try teaching a 5 year-old how to spell and pronounce the name! Which leads me to my approach to leadership.

In an effort to corral my wanderings when I was a child, my parents built a sandbox. Mom had two simple ground rules. As long as I remained in the sandbox and played there, she wouldn't bother me. Leave the sandbox, and I would immediately come to her attention. The limits of play were clear—the sides of the sandbox. So was the freedom to play in the sandbox and the consequences if I stepped outside of it.

Boundaries and freedom. Easy to say, but difficult to identify, quantify, and measure, especially for troops on position. The 1x12s that Dad used to build the sandbox could not be used as boundaries here, so another form of parameters for this sandbox called Sinop has to be defined. There are four distinct personal characteristics that I look for in each of you, and that I use to define my "sandbox."

The first attribute is competency. I will assume that you are competent to perform your job until proven otherwise. My personal experience tells me that it is usually the individual who is first aware of his or her inadequacy for the job. So my initial guidance is to let all of you know that I expect you to be competent in your job performance, and that I support all efforts to further your education and training, especially in job-related skills.

The second attribute is integrity. In my experience in combat intelligence, where only actual conflict or confrontation can prove the correctness of our intelligence analyses, absolute integrity is mandatory. The consumer of our product, the field commander, can verify the truth of all other inputs to his battle plan <u>except</u> combat intelligence. The general can count noses to ensure that all reported personnel are actually there. He can go to the firing range and personally check the accuracy of target practice. He can count the crates of bullets, beans, or body bags to verify the supply reports. But he cannot verify what is on the other side of the hill without actual combat. So on the matter of enemy forces, disposition, and strength, he and the lives of his troops depend on the integrity of the intelligence community, singly and collectively.

I also know that regardless of one's career field or personal life, when one begins to cheat on the "little" things, it becomes increasingly difficult to draw that definite line beyond which one will not go in terms of cheating, lying, or shading the truth. So personal and professional integrity is essential.

The third attribute is self-discipline. There will always be regulations and policies that enforce discipline from above, but that type of imposed discipline does not form a boundary for <u>freedom of action</u>, which is what I am trying to establish.

Today's armed forces are different from the armed forces that I joined in 1956. The discipline for today's armed forces has to come from within. All of us, regardless of rank or position have to train, educate, lead, and manage to that end. Knowledge, abilities, skills, and attitudes are all internal. The dependability to be on time for work, to say "no" to overindulgence or slacking on the job, to keep in good health for one's age, environment, and work activities, to seek self-improvement, and to demonstrate pride in one's efforts all come from within. These are all manifestations of self-discipline, and all help contribute to that third side of the box. There is a direct link between integrity and self-discipline because it appears that one characteristic cannot exist without the other.

The bottom line—and the fourth attribute—is trust. Mutual trust. I trust that each of you will do what is expected and perhaps more when the occasion requires it. And you must trust that I will do what is right for the accomplishment of our objectives and for your welfare.

The concept of trust warrants further discussion. I maintain that included in the practice of trust is the freedom to fail. Freedom to do what? To fail? Yes. Without some degree of risk, there can be little forward movement. With risk, by definition, comes the possibility of making a mistake. However, if that mistake occurs as the result of a well-intended decision in keeping with the desired goal, and the preparation was reasonable, given the circumstances, then the mistake becomes a learning experience for all concerned. I should note that freedom to fail does not imply nor is it an excuse for incompetence, indifference, or ignorance.

Another aspect of trust is to recognize and accept friction. While an analogy never proved anything, there are learning points to be gained from the use of one. Consider that some friction is essential for a wheel to move forward. Without friction the wheel will spin in place. With too much friction the wheel will bind, and there will be no forward movement. So it is with organizational positions. There must be a certain degree of healthy friction between and among all parties if the organization is to maintain any degree of forward momentum. The mutual belief and trust that all parties are working for the same objective helps keep that friction within reasonable bounds.

A final dimension of trust is summed up in the expression "give a damn." Everyone in this theater went through basic training. During that time, maybe you were in a group that claimed it would set a record time on a distance run. And then the leader set a blistering pace, with the result that the leader and maybe two or three others made it across the line in record time. However, the rest of the group was strung out piecemeal along the road, left to fend for themselves.

On the other hand, maybe you were in a different kind of group in which the objective was to set a record time as an organization. There the members pulled together and carried, if necessary, those who couldn't keep the pace or who got injured along the way. Perhaps a record time was set, perhaps not. But the entire group finished as a group, with everyone giving all and then some.

Have you ever considered what the subtle messages of those two groups might be and what those messages have to do with giving a damn? I suggest that the first leader was saying that it was every man for himself and devil take the hindmost. The message was that you couldn't trust anyone but yourself because no one else cares. This is a poor attitude to carry into a stress-loaded environment because individuals will always be looking out for Number One and always keeping a little bit in reserve for self-preservation purposes. Why not? They have seen proof that one has to look out for oneself because the leader sure won't.

The second leader was saying, "<u>We</u> give a damn." The second group's experience illustrated that if you gave all you had and then some, but still fell short, there would be someone there to support you. That's because you too were ready to help someone who needed it. Mutual trust. And that is the message behind my slogan. <u>Be concerned for, aware of, and trustful of your team members.</u>

These are the four sides of my sandbox: competency, integrity, self-discipline, and trust. And being a perfect square, the length of each side is equal to the length of the shortest side. The shortest side determines the overall dimensions for freedom of action because the personal characteristic that is in shortest supply in a given individual will determine the overall degree of freedom of action that an individual can expect.

However, being a square, the box is an unstable structure unless there is some device that is used to hold the structure in place. In my childhood sandbox, that device was a floor. Here in Sinop, that device is communication. Without clear, two-way communication, including listening, on a regular basis, the sandbox can become rather unstable. The sides will begin to shift and possibly shrink. So communication is added as the stabilizing factor to this sandbox.

Each box is an individual one—one box per person, determined by that person to fit that person. While we may be viewed by what we do as an organization, we are still measured individually on performance appraisals, effectiveness reports, or similar grade cards. Therefore, each sandbox must have a unique, personal fit to it, since the attributes that make the sides are all personal attributes determined by one's own individual actions.

Once the overall limits are established and the floor is in place, then the individual must be given the freedom to operate within the given limits of that box. In my experience, Americans generally prefer to be told <u>what</u> to do and not <u>how</u> to do it, especially those of us involved in the technological, intellectual, operational, or leadership fields.

Once you NCOs, as the supervisors, have established the sandbox for each of your troops, and the individual has given it some personal dimensions, then back off and permit that individual freedom of action, including freedom to fail. Give them the freedom to do what has to be done. But always do so with the admonition, "When you step outside the sandbox, you will come to my attention." There is a deliberate absence of any negative term in that admonition because stepping outside the box may indicate a need for help, a need to redefine the mission, or a need to reassess the situation or the dimensions of one side of the box. Perhaps there may also be a need for you as supervisor to pay more attention to that individual on a personal basis.

The application of this method is not for the faint of heart. Consistent application requires virtually every positive attribute noted by writers in the field. These include management by walking around—MBWA. My first boss called it "being in the motor pool where your troops are and not in the mess hall where the coffee is." It includes the use of lean staffs, knowledge of one's business, and the use of loose-tight controls. It takes effort, time, and patience to bring about an environment where the sandbox approach can be used.

There are some specific issues I want to address before I open the floor to whatever questions you have of me."

At that point I left the script and explained what fraternization was, why it was harmful to an organization, and why it would not be tolerated. The example I used was of a senior NCO or junior officer bedding down with a lower-ranking troop. I let it be known that fraternization would not be tolerated—period. I kept it short, simple, and direct, and moved on to other topics, such as use of the learning center, opportunities to use the "White Boat," the wearing of BDUs in lieu of greens, and keeping in good physical shape, especially since we were going to take the annual physical fitness test as a matter of record. There were several other small details that I no longer remember. One point does remain though, and that was the prohibition of the public display of erotic posters and short-timer's calendars. More on the that later. The briefing and the Q and A session lasted just over an hour, which was my goal. Norm and Russ appeared to be satisfied with my approach and the positive response of many in my audience.

The day ended with a lengthy give-and-take session in the DOOM over freedom to fail and power down. The cynics, and there were a few—after all, this was Diogenes' hometown—left me with the impression that actions would have to speak louder than words. I had felt the same thing myself on more than one occasion, and I knew that time would tell.

Chapter Six
Talking the Talk, Walking the Walk

During the early 1980s, the U.S. Army, along with many corporations, was taken with a new concept of management development—not necessarily new management procedures, but a new way to develop and structure management to get the most effective and efficient outcome possible. This concept centered on what was referred to as "organizational effectiveness" or OE. INSCOM, my next-higher, was into OE lock, stock, and barrel. INSCOM had a team of three or four junior officers, i.e., captains and majors, who were graduates of the organizational development course at Fort Ord, California. I watched them in action while I was in the INSCOM training sessions. I thought that what they offered to new commanders was beneficial.

I had some real problems for which there were no simple or all-inclusive solutions. Fraternization ranked highest in my mind, but there were others. We had stuck around 500 young Americans on a remote spot of land—one that stands out whenever you look at a map of Turkey, regardless of the scale of the map. They were foreigners any way you looked at it. We had a "Welcome to Turkey" class for all new personnel. We covered basic language skills, such as greetings, numbers, and phrases like "how much, "thank you," "please," and "goodbye." We also covered a little history, explaining why there was the ubiquitous statue or painting of Mustafa Kemal Ataturk everywhere you went. And basic manners: never physically touch a Turk, respect the religion and its practices[1], and shopping is not just buying—it is a social event. Try as we would, we still had many GIs get on the outbound plane claiming that they never left the compound in the year they were there.

---

1. Sinop was far from the foreign tourist scene and the more liberal parts of Turkey. We were smack dab in the center of conservative Sunni Turkey.

Besides the strangeness of the area, there were the related issues of fraternization, potential alcohol abuse, and depression. Let's be honest—the site was at the top of a windswept hill, devoid of trees or anything of scenic beauty. No one knew

Field Station Sinop - Support Buildings (right), Antenna Array (center)

the language. The call to prayer five times a day and the subsequent kneeling and facing Mecca were all unsettling to a kid just out of basic training and even to a junior officer. You might sum up the problem in one word—morale. Someone later heard me say what an outstanding personal experience being the Field Station Sinop commander was and remarked that it was apparent that my side of the desk had a totally different perspective on things than the troop's side had. And that was a valid observation.

I needed help, and the OE team might be part of the solution. So, as field station commander, my first dealing with Arlington Hall Station (AHS)—INSCOM headquarters—was to request the OE team to visit us as soon as possible. Norm was well into the OE philosophy and became the project officer for their visit.

The team was composed of three or four captains, personable and well-versed in their field. While we might have been the farthest from the AHS flagpole, we were not the first command they had worked with. I had a visit with the team when they arrived. By then, some two weeks after assuming command, I was seeing things in a slightly different light. I would continue to see things differently, even the same thing, throughout my tour, simply because factors surrounding a situation would change: personalities, levels of experience, perspectives. What was difficult was keeping the original goal in mind as I surveyed the situation. This came into play while awaiting the OE team's arrival, the salient issue being the past way of doing things versus my approach.

One request I had regarding the team-building exercise was that we would not dwell on the recent history of the command. Instead, we would focus on the future, in line with the IBEX concept of working smarter, not harder. We decided to include our logistics officer, our engineer, and our operations officer, along with CSM Welker. (Navy Lieutenant Commander Clyde Lopez was offered a seat at the table. After some discussion, it was decided that this was strictly an Army affair, and he declined the offer to participate.)

The first day was spent going through the basics of a high-performing team. (The OE team provided us with material developed and published by Dr. Linda Nelson and Frank Burns in their Army publication *High Performance Organizations*. The basic thrust was summarized in their high performance grid, reproduced below.)

HIGH PERFORMANCE GRID

	REACTIVE	RESPONSIVE	PROACTIVE	HIGH PERFORMANCE
TIME FRAME	PAST	PRESENT	FUTURE	FLOW
FOCUS	DIFFUSED	OUTPUT	RESULTS	EXCELLENCE
PLANNING	JUSTIFICATION	ACTIVITY	STRATEGY	EVOLUTION
CHANGE MODE	PUNITIVE	ADAPTIVE	PLANNED	PROGRAMMED
MANAGEMENT	FIX BLAME	COORDINATION	ALIGNMENT	NAVIGATION
STRUCTURE	FRAGMENTED	HIERARCHY	MATRIX	NETWORKS
PERSPECTIVE	SELF	TEAM	ORGANIZATION	CULTURE
MOTIVATION	AVOID PAIN	REWARDS	CONTRIBUTION	ACTUALIZATION
DEVELOPMENT	SURVIVAL	COHESION	ATTUNEMENT	TRANSFORMATION
COMMUNICATION	FORCE FEED	FEEDBACK	FEED FORWARD	FEED THROUGH
LEADERSHIP	ENFORCING	COACHING	PURPOSING	EMPOWERING

The basic High Performance Grid from the Army
publication <u>High Performance Organizations.</u>

The OE team walked us through the High Performance Grid, recognizing that there were many terms whose exact definition we were not acquainted with. The end result had two aspects. The first conclusion was that although Field Station Sinop still retained aspects of the Reactive Mode, reflected in red in the diagram, this was primarily in terms of focus, structure, perspective, communication, and leadership. Orange indicates the portions of the site where conditions were already being noticed, so rather than focus on the past, we decided to move ahead and focus on the responsive mode.

The second conclusion was that site leadership was a touchy area. The first evening that the OE team was in the DOOM, several officers walked in wearing tee shirts emblazoned with the words "I Survived the Stone Age." I asked to have one and was told I was not eligible. The tee shirt was only awarded to those who spent thirty days or more with the previous command, so I did not qualify. I read this as an area to avoid in the follow-on discussions because it was clear that we could get wrapped up in the process of venting over history we could not change.

HIGH PERFORMANCE GRID

	REACTIVE	RESPONSIVE	PROACTIVE	HIGH PERFORMANCE
TIME FRAME	PAST	PRESENT	FUTURE	FLOW
FOCUS	DIFFUSED	OUTPUT	RESULTS	EXCELLENCE
PLANNING	JUSTIFICATION	ACTIVITY	STRATEGY	EVOLUTION
CHANGE MODE	PUNITIVE	ADAPTIVE	PLANNED	PROGRAMMED
MANAGEMENT	FIX BLAME	COORDINATION	ALIGNMENT	NAVIGATION
STRUCTURE	FRAGMENTED	HIERARCHY	MATRIX	NETWORKS
PERSPECTIVE	SELF	TEAM	ORGANIZATION	CULTURE
MOTIVATION	AVOID PAIN	REWARDS	CONTRIBUTION	ACTUALIZATION
DEVELOPMENT	SURVIVAL	COHESION	ATTUNEMENT	TRANSFORMATION
COMMUNICATION	FORCE FEED	FEEDBACK	FEED FORWARD	FEED THROUGH
LEADERSHIP	ENFORCING	COACHING	PURPOSING	EMPOWERING

The Resulting grid after working with the OE team.

The primary objective now was to develop a goal for Field Station Sinop (one in which the Navy personnel could also see a role for themselves), a goal that incorporated subordinate objectives that, if attained, would assure the achievement of the goal.

Within the electronics communication field that we were part of, an annual award was presented to the most outstanding unit. To the best of our knowledge after a quick review of the historical files, Field Station Sinop had never even been a runner-up, let alone an award winner. Thus, the visible objective was for Field Station Sinop to attain the Travis Trophy.

A couple of sessions of wordsmithing resulted in the poster on the page 46. This poster summed up several key points, ranging from "Give a Damn" and "The Best There Is" to specific actions that each individual could personalize to his or her own situation.

Another awareness we shared was of the one-year tour we all had. While this was understandable in terms of troop environment, it would be a problem to pass on what we were achieving at the present time. It could be summed up in recognizing that one year from the day we developed the goal and objectives, not one of us in that room would be there. It was essentially a question of how to create a positive culture and pass it on to those who came after us.

We agreed that we would worry about that later. Now was the time to start working on the objectives, keeping the ultimate goal in mind. The foundation had been laid; now it was time to build the trophy case!

DIOGENES STATION GOAL

**To produce accurate, complete, reliable and timely results
with emphasis on pride in oneself and
Diogenes Station while planning for the future...**

OBJECTIVES

**Produce a quality product
Communicate goals, objectives, and core values
Maximize resources
Foster personal and professional development
Improve the quality of life, and
Build a sense of community**

DIOGENES STATION POSTER 1 (11 Aug. 83)

Part II

Time-Constrained Efforts at Transformational Leadership in Close and Confined Spaces

Chapter Seven
First Quarter of the Tour

Putting Theory Into Practice

When I started my 1983-84 tour in Sinop, I had 23 years of experience, including four years of officer schooling: one year each of basic and advanced officer training, a year at the U.S. Army Command and Staff College, and a year at the U.S. Naval War College. In addition, I had one year of language training at the Defense Language Institute and a two-year correspondence course with the U.S. Army War College.

All these courses, with the exception of the language course, addressed leadership to some degree, mostly in an anecdotal fashion. My own experience at leadership had consisted of one year as an airborne armored-cavalry platoon leader, one year as an airborne armored-cavalry troop commander, and eighteen months as a battalion commander at the Intelligence School at Fort Devens. The remaining twelve years were spent in staff positions, ranging from battalion level up to the Office of the Chief of Staff, graduate school, instructor assignments, and research and analysis duties.

During this same period of time—1960 to 1983—the Army had transitioned from the mind-set of the Korean War experience through some twelve years of combat in Vietnam[1], through the virtual disintegration of the military resulting from those twelve years, to the rebuilding process in 1983. The level to which the U.S. Army had fallen during that period is epitomized in the first sentence

---

1. Some say it was not twelve years of experience but rather one year of experience repeated twelve times.

of the 1976 edition of Field Manual 100-5 Operations: "The U.S. Army must be convinced it can win."

To be on active duty in a command position at a remote, strategic site gave me an exciting opportunity to try new approaches to leadership and management. Management is included for a reason: while leadership is difficult to quantify, management is not. When one is two hours by air from the nearest U.S. facility and responsible for some 500 troops, leadership and management—planning, organizing, staffing, decision-making, coordinating, resourcing, budgeting and training—go hand in hand. Leadership cannot be emphasized at the expense of management, and management cannot be emphasized at the expense of leadership.

This relationship between management and leadership has been recognized over the years and is the reason that the U.S. Army adopted the Prussian staff concept: one staff element each is dedicated to personnel issues, intelligence acquisition and analysis, operations, and logistics. Field Station Diogenes had the same staff structure, including a medical and dental staff, an engineering entity, and a legal office, all designed to assist the command in ensuring mission accomplishment. Our major concern was the one-year turn-over. You barely got to know the lay of the land before you shipped out at the end of the year. So, our effort at transformational leadership had to be simple, direct, understandable, and of such a nature that "one size fits all."

It was not until I started my doctoral efforts at the University of Maryland in 1984, working toward a Ph.D. in Higher Education Policy, Planning, and Administration with an emphasis on community colleges, that I encountered the terms "transactional" and "transformational" leadership. My dissertation was on the leadership practices of twelve community college presidents in Maryland. It was in that program, under Dr. Robert Carbone, that I learned the theories behind the practices we developed at Sinop—hence the title of this effort: A Perspective on Transformational Leadership. Usually one does the research in preparation for the job first and gains the experience later; in this case we got the experience first and learned the theories later.

I stress the plural pronoun "we." None of this would have been possible without the command team of Norm, Russ, and myself working together with the initiative and support of the field-station staff officers and senior enlisted personnel. Did everyone buy in? I seriously doubt it. Human nature doesn't work that way. There is an old saying that if everybody in the room is saying the same thing, then only one person is doing the thinking. It is always helpful to have a "red" team of one or more skeptics who can be counted on to find the holes in the logic, challenge the assumptions, and identify the shortfalls in resources. We had, I'm sure, our share of skeptics and naysayers, but we did listen and modify our planned actions on occasion, after reconsidering them based on their questions, comments and results.

The phrase "close and confined spaces" describes Sinop in the eyes of many who served there. As noted, while Sinop was not the end of the world, many Americans thought that they could see it from the Hill because of the sense of isolation. This made mission accomplishment difficult, but still feasible.

What follows is a quarter-by-quarter summary as I best remember it, based on three albums of pictures taken during those twelve months of command: August 2, 1983 to July 23,1984.

In my early days of command, I kidded Norm, asking him where he kept the famous three envelopes. He wasn't sure what I meant until I explained that when I got into trouble the first time, I was to open Number One; the letter inside said, "Blame it on your predecessor." The second time I got into trouble, I was to open Number Two. It contained a short note, "Reorganize." The third envelope read, "Prepare three envelopes."

His rebuttal was that his experience so far at Sinop had been like life in Vienna—always hopeless but never serious. This was in comparison to life in Berlin, where the situation was always serious but never hopeless. We shook hands in an agreement to keep our sense of humor, which we did for the remainder of the time we were together.

I was in command no more than a month when I received an invitation from MG Stubblebine to visit the other INSCOM sites in Europe. The objective was to begin building a relationship among the INSCOM units that would assist in the development of an intelligence group to support Seventh Army. Military Intelligence was in a building mode. We had CEWI (Combat Electronic Warfare and Intelligence) companies in support of divisions, CEWI brigades in support of corps, and now a proposed new CEWI group in support of Seventh Army.

I packed my bags, told Norm to sell if he got a good price, and took off to swing through Europe. On the negative side, it meant a week away from home station when we were just getting started. On the positive side, I would get a chance to see my son Matt, who was assigned as a junior NCO to the intelligence office of the Berlin Brigade.

Berlin was my first stop. Matt and I had a great weekend together in Berlin, especially when it came to introducing him to the dietary goodies the restaurants had to offer. His idea of a good meal at that time was a Mickey D's in the heart of Berlin, and mine was *anything* but another McDonalds! We had a good time visiting the many sights a divided Berlin had to offer. The photos show us at many locations in the U.S. portion of a divided Berlin.

From Berlin, I was off to Augsburg and then to Kaiserslautern, where I visited the proposed base for the new Military Intelligence group. From there, it was back to Seventh Army HQ in Heidelberg, where I had the opportunity to visit Ben Franklin Village in Mannheim, where Anne and I had lived from 1961 to 1963.

Ben Franklin Village 1983. The first floor apartment was ours (1961-63).

Ben Franklin Village 1963. The 1961 Valiant was ours.

The last stop was Coleman Barracks, where the 8th Cavalry was stationed, and where I served from 1960 to 1963.

Coleman Barracks - Wheeled armored car was in the same location as it was in 1960.

It was an informative trip, but I was glad to get back and see how things were going. I started the practice of briefing all newcomers the morning after they flew in on the C-12 from Istanbul. I would take the first ten to fifteen minutes to explain the "magic sandbox," give the rationale behind the "Give a Damn" slogan, and to emphasize the need for teamwork and cooperation. I stressed that there would be no public display of erotic items or short-timer calendars.[2] I explained the negative impact that such calendars have on teamwork and how, after one becomes a "single-digit midget" (99 or fewer days until departure), one's focus can be totally on the calendar, degrading the effectiveness of a team.

---

2. A countdown calendar to any date, often in the form of a pinup.

I continued the early morning farewell visits with the troops that were leaving on that day's flight to Istanbul. I would see them at the bus pickup point before they left the compound for the airstrip. I thanked them for their service and for their time at Sinop and wished them the best of luck in the future. Occasionally, someone would speak out about an issue that they had with the site management. As time went along, these complaints turned into suggestions as to how to do things more efficiently and effectively. Right up to the last day, though, we would have the occasional troop who claimed to have never left the compound. I finally resigned myself to the fact that it was just human nature, and I would have to learn to live with it.

I would swing through the barracks with the CSM and always find something out of line. That was just part and parcel of the military lifestyle in the barracks. By that time in the morning, Norm and my executive assistant, Specialist Five April Hunt,

Barracks, Field Station Sinop

would have an action or two from INSCOM HQ to respond to. My philosophy, in keeping with the "power down" concept, was that Norm, Russ, or April could take positive action on anything that came up that was in their area of influence or authority. I was the only one who could say "no" to a request. The result was fewer nit-noids at my level and more time for me to spend on relevant issues, such as field-station budgetary problems, labor relations with the Turkish work force, meetings with local civilian authorities, or even time with the Sinop museum director. (More on that point later.)

Once a month or so, I would take a tour down to our unusual power plant. It consisted of six German *unterseeboot* diesel engines. There were usually four of them turning out 60-cycle power at any given time. These German submarine engines dated back to WWII, but cruised along beautifully. The power plant and its submarine engines was always a must-see for visitors.

The power plant with WWII German submarine diesel engines.

Ironically, the 60-cycle power was a pain in the behind, and here's why: Every time we requested electrical supplies involving the word "cycles," someone at Arlington Hall Station (INSCOM HQ) would think to themselves, *"Sinop is in Europe which is 220/50 cycle power, so they must have made a mistake on the requisition form and written down 60 when they meant 50. So, I'll show initiative and send 'em 50-cycle items."* Sometimes there are drawbacks to showing initiative!

Noontime would eventually roll around, and I'd be off for my five-to-seven-mile run, from the barracks out to the operations site where the "mushrooms" were located and back. Every so often, especially when the weather was really nasty—windy, raining heavily (a mist was always there), or extremely cold—I would have lunch and "honor" one table or another with my presence—a real conversation stopper! If I didn't get to run at noon, I'd hit the small weight room for a light workout. I had initially thought I would take April's aerobic workout in which many of the women were participating. Aerobics would surely be an easy workout! I did it once and found it to be a killer. Never tried it again!

Most afternoons were spent on paperwork, meeting with the Turkish commanders on-site, or MBWA (Management by Walking Around) through a staff office. Octay Kuru had been a longtime interpreter for the site. His father was affectionately known as the NATO barber, as he was the only one in the town of Sinop who knew how to cut hair in the American style—or so the troops said.

Octay was present to fill in the necessary gaps, but most of the Turkish officers knew enough English to cover whatever needed to be covered without too much difficulty. The same situation existed with the Turkish labor force. However, in that case, we had American contractor supervisory personnel who responded to me, although I think their real chain was through the contractor support team in Europe. I was never sure, but as we had no problems, I stuck to my approach: If it ain't broke,

Octay Kuru and Bill Hanne

don't fix it. The contractors put me onto some interesting books that were little-known outside the Turkish work force: *Scotch and Holy Water* and *Tales of Nasr-ed-din Khoja*, the latter a compilation of folk stories and sayings. No matter where I went or what office I visited, I always learned something, whether a new fact, or something practical, or both.

The day closed with supper in the dining facility, either before or after a quick visit to the DOOM, followed by writing a letter home to Anne that would go out on the next flight.

It may sound woefully the same, day in and day out, but it wasn't. There was always something new going on or about to start. One ongoing activity was catching up on what someone had started years before I arrived on scene: Brian's Sail Boat. This was an example of how the one-year tour could catch you by surprise.

Our logistics office got a call one afternoon well into my tour, sometime in April or May. It seemed that the U.S. Air Force C-130 logistics run for the day had dropped off a series of boxes, a long mast, and the hull of a sailboat. The paperwork was valid, the sailboat was properly documented as being ordered by, and delivered to, Field Station Sinop. That was all. There was no indication of who ordered it, paid for it, or was in any way responsible for it. But there we were with a 20-foot sailboat on our hands at the airstrip.

Norm said something about a hand-painted mural in the operations center—a mural with a sailboat in it. I vaguely remembered seeing the boat in the mural, but I had never paid much attention to it. Up we went to the operations center to take a closer at this mural. It was some 20-feet long and maybe 10-feet high. Sure enough, centered in the mural—which was of Sinop Bay—was a sailboat with the name plate on the bow carrying the words "Brian's Boat!"

Okay, now we were beginning to tie things together. The boat in the mural matched the boat on the airstrip. We started to dig back through the unit fund files and the operations records and there it was! About five years before my tour, most likely at the end of the embargo over the Cyprus issue, an operations officer had apparently talked the unit fund manager in Europe into providing us with a sailboat out of unit funds for sports and entertainment, all legal and above board. Surprisingly, I knew the "Brian" involved; he had been one of my officers when I was the G2 of the 25th Division in Hawaii in 1976-1977. Brian was from Massachusetts and a real seawater enthusiast. So the circle was finally closed— except that Brian had long departed, and his dreams lived on through the bureaucracy of "the system."

Shortly after I returned from the "INSCOM Grand Tour of Europe," Russ approached me with a suggestion from the NCOs. Looking for something to liven up the installation, they suggested a "Dining In," a traditional evening of formal uniforms, toasts, libations, menus and jocular humor, a sort of tradition that had started with the Brits, but was carried on by the U.S. Army. The purpose was to formally recognize the camaraderie within the unit. My feeling was that it was their idea, their initiative and their action. It was a suggestion that should bring the unit together and demonstrate that the "power down" concept might be taking hold. The delegation of authority had already been demonstrated by a number of staff and operational sections, but this was a unit-wide initiative! I said to Russ, "Go for it."

The Dining In hosted by the senior NCOs was a great success, demonstrating both initiative and a sense of humor throughout the evening. There were numerous barbs directed at me and more at Norm, as he had fallen into the "black hat" position on the staff and among the troops.

*Mixing the traditional U.S. Army Field Station Sinop
punch at The Dining In hosted by the senior NCOs*

Ladies and Gentlemen it is now time to mix the traditional US Army Field Station Sinop punch.

The base for this punch, like the mission of this Field Station, and indeed of INSCOM itself, is shrouded in secrecy, known to very few and passed along very carefully.

A special batch of this base was obtained from its secret storage area somewhere under the watchful eye of the Sphinx and was jumped in during the dead of the night by a highly trained paratrooper.

When the proper ingredients are added to this base, the resulting mixture will allow you to truly understand where the lightning bolt and flaming torch on your shoulder patch come from.

The first ingredient of this punch is RAKI. It serves as a symbol as does the crescent and star on our unit crest of our ties too, and our sharing with, our host nation and it's people. For without these ties and this sharing, this Field Station would not exist.

The next ingredient is a fine BLENDED WHISKEY, which like the quartered field of our patch represents the blend of people and talents that make up our community. Without this blend we could not be so successful in accomplishing our very important mission.

The next ingredient is RUM, the traditional drink of our sister service, the Navy, with whom we share so much here.

The next ingredients are COGNAC, SCOTCH, AND GERMAN WHITE WINE, which are added as a token of our friendship to and ties with EUROPE.

The final ingredient is CHAMPAGNE - THE BEST THERE IS FOR THE BEST THERE IS.

MR. PRESIDENT WOULD YOU DO THE HONOR OF HAVING THE FIRST GLASS OF US ARMY FIELD STATION SINOP PUNCH

Stubblebine's caution about being too easy on the troops began to make more sense as I compared Norm's reaction to incidents to my own. I had the final word, but he and I both realized that the "black hat/white hat" arrangement worked for the best. And Russ was working the same routine with the enlisted personnel. Essentially, we embodied Machiavelli's observation that it is far better for "the Prince" to be feared than loved, but only if there is a "white hat" around.

Overall the power-down approach appeared to be working. Fewer issues were elevated to my level, and overall productivity appeared to be increasing. While one can quantify issues and measure one month's level to the previous month's, productivity is far more subjective. We could only go by complaints about our reports, so we had to rely on what we observed when visiting the various locations, while attempting to read between the lines. At best, we could tell that the first three months showed no major problems or even minor issues with our next higher headquarters.

Outside the barbed wire fence that surrounded the facility, things seemed to be going well. There had been no indication that the American commander was expected to attend Turkish national celebrations and holidays. We would see *askerler* (Turkish soldiers) downtown, so we knew that there was at least one other military installation in the area, as *askerler* were not soldiers from our base. Where they were stationed or how many there were was of no interest to us; we had more than enough on our plate than to get involved in something outside our designated communications activities.

Colonel Senal frequently asked me to accompany him to parades and ceremonies. I found them interesting. School children would often be in the parades, especially on national holidays. One feature always captured my attention. Without fail, in any group of fifty or more children in the parade, there would be at least one, if not two or more, fair-skinned, blue-eyed children with red hair. My silent joke, which I kept to myself, was that the Crusaders' heritage lived on! Was this a manifestation of the Law of Unintended Consequences?

Octay's wife, whom I met only two or three times, was teaching the sixth grade in the local school. Octay mentioned once that she would be teaching first grade the next year. In response to my surprised expression, he explained that she had been this group's teacher since the first grade. In other words, the teacher would start with a first-grade group and stay with them through the sixth grade, after which they evidently continued their schooling or went into a vocational trade. Once again, I was not in Turkey as a sociologist, but as a field-station commander, so while there were many items I wanted to know more about, the opportunities for doing so were limited.

A snapshot of how we thought we were doing is presented below. It is well-known that just a simple change in procedure or approach has a bump effect for a while, and that is how we assessed the situation, reflective of a three-month change in leadership and focus. Would it last?   That was the question.

HIGH PERFORMANCE GRID

	REACTIVE	RESPONSIVE	PROACTIVE	HIGH PERFORMANCE
TIME FRAME	PAST	PRESENT	FUTURE	FLOW
FOCUS	DIFFUSED	OUTPUT	RESULTS	EXCELLENCE
PLANNING	JUSTIFICATION	ACTIVITY	STRATEGY	EVOLUTION
CHANGE MODE	PUNITIVE	ADAPTIVE	PLANNED	PROGRAMMED
MANAGEMENT	FIX BLAME	COORDINATION	ALIGNMENT	NAVIGATION
STRUCTURE	FRAGMENTED	HIERARCHY	MATRIX	NETWORKS
PERSPECTIVE	SELF	TEAM	ORGANIZATION	CULTURE
MOTIVATION	AVOID PAIN	REWARDS	CONTRIBUTION	ACTUALIZATION
DEVELOPMENT	SURVIVAL	COHESION	ATTUNEMENT	TRANSFORMATION
COMMUNICATION	FORCE FEED	FEEDBACK	FEED FORWARD	FEED THROUGH
LEADERSHIP	ENFORCING	COACHING	PURPOSING	EMPOWERING

Chapter Eight
Second Quarter of the Tour

We had our fair share of bad weather coming off the Black Sea, but I do not remember any really severe storms during the winter of 1983-84. It's all relative, of course. Having lived through the Blizzard of 1978 in Fort Devens, the Blizzard of 1996 in D.C., and Super Storm Sandy in 2012, any storm at Field Station Sinop would have to have been a real humdinger to make a lasting impression in my memory cells! That said, there were occasional whiteouts and blizzard-like conditions when the troops used the ropes for safety and to guide them from the barracks to important places like the mail room and the dining facility.

My dropping in on operations at 3:00 a.m. was no longer a rare event. I never was sure whether the troops in ops had not worked something out with the MPs. When I used an MP vehicle to go out to operations, I wondered if the desk sergeant had alerted them that I was en route. I wasn't playing hide and seek; I was trying to ensure continuity of operations and preclude nasty-grams from the States about a lack of responsiveness on our part. MBWA seemed to work, so I continued the unannounced (or so I hoped) early morning visit through the end of my tour.

I now sensed that we had forgotten about the negative aspects of the past and were focused on the future. Good! Plus, we were entering the holiday season! Someone suggested using the gym for a Halloween Party, complete with costumes. I was scheduled to be in Ankara at that time, so I tossed the overall responsibility into Norm's lap and let him run with it. When I got back to Sinop a couple of days later, there was still a lot of kidding and comments around the dining facility about how much fun the party had been. Apparently, the costume contest had been won by two female staff sergeants who went as a pair of M&Ms, complete with the M&M slogan about not melting in your hand.

What was more interesting is that about this time, we began to see a downturn in liquor sales. As I recall, there had not been any need for non-judicial punishment activity related to being late for duty or similar offenses. I'm sure that in many cases of minor violations, the senior NCOs who were team chiefs or section chiefs worked things out on their own. It was usually the kid who had been given several chances but failed to respond who ended up in front of my desk, and that had not happened… yet.

I don't recall anything special about Thanksgiving Dinner, but I'm confident that the traditional full spread was there. Our Turkish cooks had been at the base for some time. And I'm sure there was at least one who had been there since the site opened in the late 1950s. The impact of the dining facility (a.k.a. the "mess hall") was significant. As I remarked earlier, the GI's morale rested on a three-legged stool—all legs beginning with an "m." The three legs were mail, mess, and money. You did not fool around with any of those three "m's," or you had a major morale problem on your hands.

Lieutenant Commander Lopez and I had a bet on the Army-Navy game, all part of the professional competition that existed between the two teams. Unfortunately, Army lost the game, and I ended up riding an *esek* (Turkish for "donkey") backwards around the patio of the DOOM—good laughs all around. We were all familiar with the *esek*; they could be seen in many a farmyard and pasture around Sinop. Our own fleet of two C-12s was known as Esek Airlines.

Riding the esek (donkey)

About this time I picked up an additional role: disk jockey on our local radio station. We had—remember, this was 1983—plenty of vinyl (33 1/3 rpm records) to choose from in our library. I focused on semi-classical, which I spun for a couple of hours on Sunday afternoons. Our radio broadcasting was actually a wire operation through the internal electrical system. We were across the Black Sea from the Crimean Peninsula, and the Turks were not exactly appreciative of our music and foreign ways, so if you plugged in a radio you got our transmission via the wiring in

One of the fleet of two C-12s known as Esek Airlines.

the buildings. My theme was Glinka's *Overture to Ruslan and Lyudmila,* perfect for our site.

I did not get many—let's change that to *any*—requests for changes in my choice of music. It was my way of getting to listen to music I enjoyed under the pretense of doing my share of community assistance—not overly subtle but, as I put it, anyone who wanted to volunteer from the hours of 2:00 till 4:00 on Sunday afternoon was welcome to my spot. There was never a volunteer. There's more than one way to skin a cat…

Later on, during the basketball season, I called a few games on the radio when I had the chance. I found out later that my volunteer hours earned me a slot on the radio's softball team in the spring.

DJ Hanne

With Halloween, Thanksgiving, and the Army-Navy game past history, the troops were now looking ahead to Christmas. We tried to be as liberal as possible in terms of leave time back in the States, but mission requirements had to come first. As I recall, we tried working 12-hour shifts during the week before Christmas, over the holiday period, and for a week after New Year's, almost a month in total. We found a few gaps in the system, which we were able to cover, but also tested our limits of endurance while maintaining required capabilities. I stayed at Sinop over the holidays, so Norm could be gone. In return, he would be available to cover for me at the end of January.

The Chapel

Father Carrol Thorne, our Catholic chaplain, held the traditional Midnight Mass in our chapel. After the service, I walked out to the crest of the hill overlooking the Black Sea. I wondered if at the time of the Gospel, when it was required that all go to their home for the purpose of the census, a Roman soldier might have stood where I was standing, looking out over the same body of water. My memory of a moonlit night, fairly calm and quite peaceful is rather wonderful.

Fathers Thorne and Grant

However, when Father Thorne and I went down to the airstrip to give the flight crews and the guards a holiday greeting, it was a different story. As we crossed the tarmac to the hangar from the shadows, we heard a distinct *click-click* as a round was chambered. We stopped and hurriedly turned the flashlights we were carrying so they lit up the two of us. Then the guards knew who we were. I assure you, once you hear a round being chambered, you never forget what that sound is.

We came back up to the DOOM, where celebrations were moving along at full speed. Somehow, everyone got some kind of gag gift. I've forgotten who put it all together, but someone showed initiative and team spirit and helped those of us who were separated from our families enjoy the new "family" we had at Sinop. I remember that the favorite drink that evening was Irish Coffee served from gallon-sized containers! We promised the stand-by flight crew, who could not drink, that we'd save them some to enjoy when they came off shift.

Christmas celebration in the DOOM

L-R: CW4 Warren George, Pilot
Lt. Vic Irwin, FAO
Cpt Mike Bodman, DFEH asst
Cpt Bill Dowling, DFEH
Maj Hank Scarangella, FSO

In a heartbeat, it seemed, 1983 was out the door, and we were focusing on writing *1984* in lieu of *1983*. We all agreed that this 1984 had best not be like George Orwell's novel! Routine set in, and the biggest struggle was to keep the troops motivated to continue in the patterns and work habits they had formed in the past few months.

Christmas celebration in the DOOM
L-R: Rebecca Lustig, Tom Neal

Looking ahead at personnel projections, Norm, Russ and I were bothered by a projected decrease in an operational specialty. We usually got an official communication listing who was inbound and their projected arrival date, but February and March were looking rather empty. If we didn't start getting personnel fills, we would have to move to 12-hour shifts with a day off every ten days. That routine would kick in sometime towards the end of January, about three weeks away.

Russ took care of notifying the team leaders and senior NCOs of the forthcoming problem, while Norm took over working with personnel back in the States. I wanted it that way—people at the appropriate levels doing the work, while I kept an invisible tight handle on things. There was no sense in my getting involved now with INSCOM; Norm and Russ were making the contacts and raising the awareness.

About this time, several matters that directly affected us came to a head. The primary one was access to fresh water. A related issue was a pending confrontation with the Turkish labor unions whose members worked on the installation in a number of capacities, ranging from cooks to power-generation personnel that kept those eight submarine generators going. The American contract supervisors in both areas kept us informed and asked us to stay out of the way. They said HQ in Europe was aware of the issues, and they were working them.

Water had been a long-standing issue. Water rates kept rising, so a couple of years before my tour, a desalination plant had been started that would supply us with fresh water within a month or two. It was interesting that the Turkish authorities had been watching the construction for several years and hadn't said anything. Then, about three months before the plant was scheduled to go into operation, the Turkish authorities stated their concerns and wanted us to stop. It was too late. The plant came online, and the morning coffee visits became a little tense as Sinop lost a source of revenue with the construction of our plant. Europe's advice was followed, and all went well from our side. The same with the labor issues—I did not ask questions. This was being handled at pay grades several levels higher than mine, a condition I was most thankful for.

Then a major medical problem erupted when our lawyer, Charles Pearce, had a brain embolism while he and I were lifting weights together. Our stand-by crew was ready to go, and on the advice of Doc Snyder, our physician, we flew Captain Pearce to Landstuhl Army Hospital. It was a scary situation: I had no EMT training, and I didn't know what to do when Charles started having convulsions. He survived, for which we were all thankful, especially since he had a quiet sense of humor and was great at defusing tense personal situations.

Octay, our interpreter, approached me one afternoon and asked if I would like to visit the museum. I eagerly accepted his offer. That Saturday, the museum director, with Octay's help, took me on an extensive tour of the museum, including visits to key points in Sinop's walls—walls that were started by the Greeks, added to by the Romans, and finished off by the Ottoman Empire. The director and I struck up an immediate friendship, and out of this simple tour, I was able to expand the awareness of our troops of the history of the area. But there was a catch—there always is. The director wanted to know if arrangements could be worked out where he could "borrow" some of our engineering equipment and operators to do some heavy lifting at the museum.

Sinop Museum

I checked with Bill Dowling, our engineer. After a few questions, we determined that the equipment the museum director needed saw little use on our installation. The operators for the museum work would be the same people who used the equipment on post. Plus, the work off-post would be on weekends when the equipment would not be used except in an emergency. By pushing the envelope a little, we worked things out so that any heavy lifting that the museum needed would be handled by our people using our equipment in a civic action process. It worked out like a charm. And we eventually evened the debt out to our benefit, as you will see later.

Sinop Sea Martyrs Memorial

My time as commander was half over, and while we held weekly sessions with the managers and chiefs of various operational and staff segments, we needed to step back and see where progress had been made in our efforts to become a high-performing organization. Half of the officers and senior NCOs who were onboard when I took over in August had rotated home and had been replaced one at a time by newcomers. (I should note that there was a small percentage of troops who were on repeat assignments to Sinop, always at their own choice. These were usually involved in some challenging tasks, sort of "one-of-a-kind" jobs, especially in the electronics maintenance sector. However, these jobs were very few in number.) Most of the troops who arrived were full of the old wives' tales about Sinop—isolation, reclusiveness, instant and easy access to alcohol, suicides—whatever was negative about the site. Many of the replacements had heard all of them. So we had a re-education process to go through. This effort, thankfully, was handled by the senior NCOs on-site and on position.

As we developed as a functioning staff, we encountered the task of establishing priorities among a number of options. Nothing was ever a no-brainer in deciding which was top priority. Frequently, the need for establishing a priority among competing options was financial. Could we afford it, whatever "it" was? Norm introduced an approach which essentially confirmed that it was easier to establish a priority between *two* choices than establishing a priority among *eight* or more. We immediately labeled the method as "Chung's Choice" and used it quite often to assist us in our decision-making.

I found Chung's Choice useful in my personal life. The example that follows helped me establish priorities of places to consider for retirement. I was looking ahead! Anne and I had eight possible sites, all of which we really enjoyed and strongly considered as retirement spots that we should check out as the opportunities presented themselves. But every time we started to get serious, we'd get lost. There were too many options! Norm's method helped us address that issue in a simple manner.

Step one was to list the options and assign to each a sequential letter — A through H in this example. The second step was to decide between the two options presented in each pairing. It was easy. Just stick to selecting which one of the two presented was your top choice. The choice would be between just two in every instance, with the winning selection being noted in bold and red in this example.

Location	Alphabetical	Dual Choice						
Arizona	A							
New York	B	**A**B						
Maryland	C	**A**C	B**C**					
Pennsylvania	D	A**D**	B**D**	C**D**				
California	E	**A**E	B**E**	**C**E	**D**E			
Kansas	F	**A**F	B**F**	**C**F	**D**F	**E**F		
Hawaii	G	**A**G	B**G**	**C**G	D**G**	**E**G	**F**G	
Virginia	H	A**H**	B**H**	**C**H	**D**H	E**H**	F**H**	G**H**

TOTAL SELECTIONS								
A-5	B-0	C-4	D-3	E-2	F-1	G-0		A-5
B-0	C-1	D-1	E-0	F-0	G-0	H-1		B-0
C-0	D-1	E-0	F-0	G-0	H-1			C-5
D-1	E-1	F-0	G-1	H-1				D-6
E-0	F-1	G-0	H-0					E-3
F-0	G-1	H-0						F-2
G-0	H-1							G-2
H-1								H-5

The third step was to determine by individual letter the number of times that option was the top choice. The fourth and final step was to determine which choice was selected as the top choice. The eight choices were narrowed down to one—D. It was far easier to choose between each pairing than among all eight. It is possible to end up with a tie between two or more choices for top selection. However, the choice is still always a choice between two alternatives, rather than eight.

To see if our efforts had been worthwhile, we held a short assessment session with many of the original cast, plus some of the newcomers. I stayed back from the discussions, observing and—for once—keeping my mouth shut, especially when it came to the reason that something happened. If I believed in the concepts of "power down" and "freedom to fail," I had to "walk the walk."

The leadership team came up with the following ratings, indicating progress in many areas. Overall, we had progressed from Reactive to Responsive mode, with several outliers in the Proactive mode. While no one felt we had regressed in any of the areas, there were several where progress was not as good as it could have been.

HIGH PERFORMANCE GRID

	REACTIVE	RESPONSIVE	PROACTIVE	HIGH PERFORMANCE
TIME FRAME	PAST	PRESENT	FUTURE	FLOW
FOCUS	DIFFUSED	OUTPUT	RESULTS	EXCELLENCE
PLANNING	JUSTIFICATION	ACTIVITY	STRATEGY	EVOLUTION
CHANGE MODE	PUNITIVE	ADAPTIVE	PLANNED	PROGRAMMED
MANAGEMENT	FIX BLAME	COORDINATION	ALIGNMENT	NAVIGATION
STRUCTURE	FRAGMENTED	HIERARCHY	MATRIX	NETWORKS
PERSPECTIVE	SELF	TEAM	ORGANIZATION	CULTURE
MOTIVATION	AVOID PAIN	REWARDS	CONTRIBUTION	ACTUALIZATION
DEVELOPMENT	SURVIVAL	COHESION	ATTUNEMENT	TRANSFORMATION
COMMUNICATION	FORCE FEED	FEEDBACK	FEED FORWARD	FEED THROUGH
LEADERSHIP	ENFORCING	COACHING	PURPOSING	EMPOWERING

One of my mistakes involved the communication factor. Every so often, the communications system would send out a world-wide "check-in," a sort of critical outlook at responsiveness. One morning Norm came in and shut the door. That was never a good sign. It turned out that we had "ignored" this check-in call. Of all the systems involved, we were the only one that did not respond. My guidance to Norm was twofold: if we heard from our next higher-up, I would take the blame, and we would ensure that it would not happen again. To be proactive on the latter, I had Norm get together with the key NCOs and officers and determine what had happened and why. It turned out to be a lack of training and education on our part, based largely on the assumption that all of our operators were experienced with these critical messages. And as with all "assumes," it makes an "*ass* out of *you* and *me*."

An all-too-quick trip home in late January ended my first six months in command, though I felt like I was just getting started!

Chapter Nine

Third Quarter of the Tour

Addressing the prospect of a personnel shortage, as mentioned in Chapter 8, was a priority as I began the third quarter of my tour. LTC Norm Chung and CSM Russ Welker had been working on the issue, but the problem was far from being solved.

Personnel status was one of the few areas I held tight because we depended on highly-trained and reliable troops to accomplish our mission. In the U.S. Army of the early 1980s, such qualified people could be hard to come by. We rotated troops out based on the orders we received from INSCOM. They knew as well as I did what the strength levels in the field station were. I would not even consider keeping a soldier past his rotation date. I even doubt that I had the authority to do so. But when I arrived back on-site after ten days in the States, we still had not received word of incoming replacements.

I asked CSM Welker to meet with the senior NCOs and work out a shift plan that would keep all positions covered on the required 24-hour basis. I relied on Norm, who had served in field stations before and was more familiar with procedures than I was, for suggestions on what was feasible.

We tried all combinations, ranging all the way up to 12-hour shifts for 30 days without a break. I was now in contact with INSCOM personnel almost daily. I received no satisfaction, so I finally asked Norm what I could do that would get their attention. His reply was direct: close a position and continue to do so until personnel levels permit a relief.

My next call was to INSCOM operations. I was finished with personnel—they offered no assistance. I told the operations chief what I was doing. I was told that I could not do that—period. I simply said, "Try and stop me." My rationale was that the troops were just about at the end of efficient operations, and that was the major concern. It would be better to operate four positions, for example, at full capability, than five with half-assed incompetence. The conversation was short, direct, and bitter. Norm and Russ were present, listening to my end. There was dead silence in the room when I hung up. They both agreed that they had never heard of a similar action, but they also had not been in a station where the personnel situation had been so grim.

Within two hours (as I recall), there was a terse message from INSCOM: There will be a number of troops in Istanbul tomorrow morning for pickup and transfer to Sinop. I'm sure these three troops were not in the best of humor. They were coming from the high life of West Berlin to the foggy shores of the Black Sea, but that was not my concern. We did not have to close a position as I had threatened, and we had no personnel problems after that, but I learned later that I was not held in high opinion by the INSCOM staff.

Having been around Stubblebine before, I figured INSCOM ops would do their best to resolve the issue before going to the Commanding General. The INSCOM staff did the same thing I would have done—attempt to resolve a problem at the lowest possible level—power down! By the way, the troops onboard the C-12 the next afternoon were indeed one dismal group as they looked around, comparing the Hilltop enlisted club and our immediate environs with what they had said *auf wiedersehen* to yesterday. But such is life. I never heard another thing about my alleged "blackmail" (yes, that term was kicked around, or so I was told by some who were in on the planning at Arlington Hall), and I chose to let sleeping dogs lie.

The Berlin additions remained with me up to either the time I left or their tours in the Army were finished. They were good, experienced troops. Once they saw what we were dealing with in terms of personnel levels, there were no complaints—at least I did not hear any. But I did remember what that one NCO said—the view from my side of the desk was often quite different from the view from the other side.

There were a few rough edges to the use of the "power down" and "freedom to fail" objectives. The budget was one of those rough edges, largely because we did not do a good job of keeping the details front and center. This was traced back to a lack of history. If you can't see where you came from and how you got there, it can be hard to see where you are going. Remember "Brian's Boat" and what we had to go through before we realized it was ours and already paid for—from the U.S. end. There might have been some issues from the Turkish side, but my tour was up before all the paperwork was finished.

Some thirty years after leaving Sinop, I now realize that Sinop was a great deal like Vietnam. We started Sinop in 1956, and here we were in 1984, almost thirty years later. But it wasn't thirty years of experience (except in some relatively small, but significant areas). It was one year repeated thirty times. Some of the issues caught us off-guard. Take the plumbing…please!

Sinop water was heavily calcified, especially in the water pipes that carried hot water to the radiators in the barracks. The details escape me, but the story goes like this. In a galaxy far, far away (Washington, D.C.), in a time long ago (before 1976), someone thought ahead and put re-plumbing into the projected financial plan, part of the ill-famed Planning, Budgeting and Programming System. The system is a logical one, as was almost everything that came from McNamara and his Whiz Kids, but the Law of Unintended Consequences that led to Vietnam also led me to almost overrunning our budget.

You determine there is a need. For example, the weather on the north coast of Turkey, fronting on the Black Sea, requires heated quarters if troops are to survive the winter. Therefore, you must plan for hot-water heating and plan for a replacement cycle as well. This was originally implemented by the U.S. Army Security Agency — "Massachusetts Power and Light" (MP&L) in Channon's cartoon. The next step is to budget for it, usually over a five-to seven-year period. Once the budget for the project is approved, it goes into the Program phase, where it becomes part of the Army's fiscal expenditures for the coming year(s). Makes sense, right?

Channon's cartoon. (full size on page 12) *"Massachusetts Power and Light" circled.*

Between the time MP&L planned for the replacement and 1984, when the project engineers showed up unannounced on my doorstep, MP&L became part of INSCOM and over-obligation became a capital sin. (Translated, that means that if you sign a contract for which there are no funds specifically earmarked for said project, you get fired—and it does happen. I saw the principal budget officer for the Army, a Lieutenant General, get the axe because he did what we always used to do—only the law had been changed, and his staff was unaware of the consequences.)

But replacing the pipes was only one contract that was fulfilled. The second was a stuccoing of the outside walls of the barracks. Once again, this was designed to save energy and keep the troops warm. (Norm had departed the scene as this drama came to a close, replaced by another very capable deputy commander, Jimmy Webster.) Neither project was in my budget, nor could we find any evidence whatsoever of paperwork at our end. Fortunately, Arlington Hall Station could track things back and was able to fund both projects. Nevertheless, we were frustrated because while we literally had nothing to do with it, we could have done a better job tracking and preparing for the renovations if we had known about the contracts. It shows the importance of communications!

I've never been one to pass up an opportunity to see new terrain, and much to my delight, I got to do some traveling in Turkey during the winter. Father Thorne, who had celebrated our Christmas Mass, was the only Catholic chaplain for the U.S. Army in Turkey at the time. There were several artillery installations and ammunition storage facilities elsewhere in Turkey, so Father Thorne would make circuit-rider trips to visit them until another chaplain was in place. I'd grab a seat on the C-12 and accompany him on his trips. That way I got to see a great deal of eastern Turkey, including some good views of Mount Ararat (5,137 meters or 16,854 feet in elevation). We centered our visits on the Erzurum region. On some C-130 flights out-and-back to Sinop, I also visited Diyarbakir and Izmir. On one trip we did something only a geographer would appreciate: we flew over the Meander River.

It was in Erzurum that I first saw what is now known as the Anatolian Shepherd dog, a massive beast and a protective one to boot. The heritage and genetics of this animal would be interesting to investigate, as historians note that massive animals accompanied Genghis Khan's Mongols when they moved west across the Anatolian plateau. I can imagine the scene— Mongol warriors moving right along with packs of Anatolian Shepherds.

Anatolian Shepherd

I also took a couple of bus trips. One was with the basketball team when they played a Turkish team. I do not remember whether we won or not, but I do remember that I froze my butt off in the hotel room that night. The heat went off at nine and did not come back on until after we checked out. But Octay made up for it by showing me through the ancient city of Amasya, where the cliffs that line the south bank of the river are honeycombed with massive caves, all hand-hewn from

Kral Kaya tombs in Amasya

the natural stone. I found out later that the caves are known as the Kral Kaya tombs. From what Octay told me, they date back to Hittite times.

I asked Octay why so many apartment buildings were in an unfinished stage—rebar still exposed, top floor not walled in, and other features incomplete. He told me that the taxes on a building still being built were considerably lower than on a completed building, so many people never "finished" their residence.

My second bus trip turned out to have an unwelcome side effect. I had to make a trip to Ankara on business, so I flew down on the regular morning flight. Because waiting for me would have delayed the return flight from Istanbul and because the plane was full of new arrivals, I took the alternate means of transportation back to Sinop—a bus.

I had grabbed civilian clothes before I left. When we landed, I had Mike Lustig drop me off in Ankara where the attaché met me and took me to the meeting. I caught the bus the next morning. It was one of those million-dollar trips you wouldn't give a dime for! The bus trip turned out to be a milk run, and it took me all day and well into the evening to make it back to the installation. I had been cautioned about not eating milk products, but I knew better than our surgeon and had some yogurt about two hours out of Sinop. Thank God it wasn't three or more hours. I had a horrible case of the GIs, which I attribute to the unpasteurized yogurt, just as I hit my room. Timing is everything. What did I learn from this experience? To read time tables, not to eat yogurt, and to stay close to a commode, even bomb sight commodes if you have to!

In March several of us traveled from Sinop through Athens to Landstuhl, Germany. There were two reasons for the trip. First, Hank Scarangella, our Operations officer, and I wanted to check on Charles Pearce and see what, if any, lasting damage had occurred from the brain hemorrhage. Second, we were scheduled to visit CEWI (Combat Electronic Warfare and Intelligence). We stopped at the hospital en route to CEWI headquarters and were allowed a brief visit with Charles. The good news was that he was virtually fully recovered with minimal aftereffects. The bad news was that he was being shipped to Walter Reed in D.C. for further observation and workups. We hated to lose him, but we were happy to know that he was on a positive path.

The stop at the CEWI group's headquarters (still under development) was interesting, but there was still too much they had to do before we could look to them as the next higher headquarters for sites such as Sinop. We would still do better with INSCOM for issues needing resolution—if they would even let me in, of course. Our C-12 was undergoing some maintenance at Rhine-Main while we were making our stops. The timing worked out nicely, and we were back in the air, with a scheduled layover in Athens which we all enjoyed. We visited all the usual tourist haunts while additional work was done on the C-12.

It had been a rough quarter—a death, holiday scheduling, personnel shortages, visitors and near-visitors, trips, and screw-ups of our own making. The death was that of a civilian contractor who died of a heart attack while servicing our equipment. As he was not totally covered by the DECA/SOFA, being a contractor and not a uniformed individual, the Turkish equivalent of the coroner had a role to play, but I left the details to be worked out with the key personnel, expressed our condolences, and all was handled smoothly.

Athens

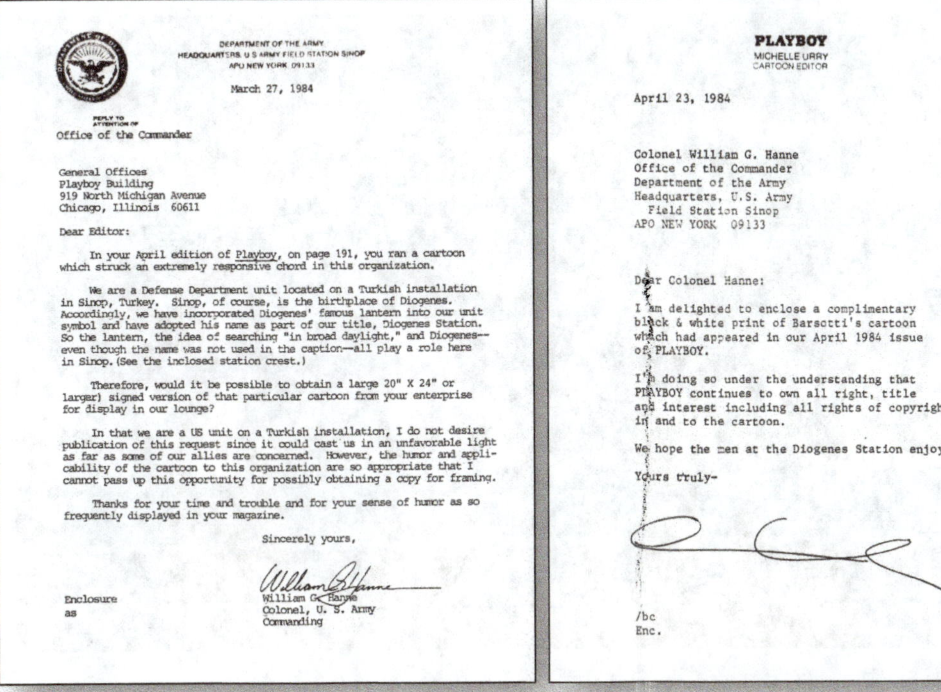

DEPARTMENT OF THE ARMY
HEADQUARTERS, U S ARMY FIELD STATION SINOP
APO NEW YORK 09133

March 27, 1984

REPLY TO
ATTENTION OF

Office of the Commander

General Offices
Playboy Building
919 North Michigan Avenue
Chicago, Illinois 60611

Dear Editor:

In your April edition of Playboy, on page 191, you ran a cartoon which struck an extremely responsive chord in this organization.

We are a Defense Department unit located on a Turkish installation in Sinop, Turkey. Sinop, of course, is the birthplace of Diogenes. Accordingly, we have incorporated Diogenes' famous lantern into our unit symbol and have adopted his name as part of our title, Diogenes Station. So the lantern, the idea of searching "in broad daylight," and Diogenes—even though the name was not used in the caption—all play a role here in Sinop. (See the inclosed station crest.)

Therefore, would it be possible to obtain a large 20" X 24" or larger) signed version of that particular cartoon from your enterprise for display in our lounge?

In that we are a US unit on a Turkish installation, I do not desire publication of this request since it could cast us in an unfavorable light as far as some of our allies are concerned. However, the humor and applicability of the cartoon to this organization are so appropriate that I cannot pass up this opportunity for possibly obtaining a copy for framing.

Thanks for your time and trouble and for your sense of humor as so frequently displayed in your magazine.

Sincerely yours,

William G. Hanne
Colonel, U. S. Army
Commanding

Enclosure
as

PLAYBOY
MICHELLE URRY
CARTOON EDITOR

April 23, 1984

Colonel William G. Hanne
Office of the Commander
Department of the Army
Headquarters, U.S. Army
  Field Station Sinop
APO NEW YORK 09133

Dear Colonel Hanne:

I am delighted to enclose a complimentary black & white print of Barsotti's cartoon which had appeared in our April 1984 issue of PLAYBOY.

I'm doing so under the understanding that PLAYBOY continues to own all right, title and interest including all rights of copyright, in and to the cartoon.

We hope the men at the Diogenes Station enjoy it.

Yours truly-

/bc
Enc.

There were humorous moments as well. Late in March, the April 1984 issue of *Playboy* arrived. In it was a cartoon of Diogenes going through the streets with his legendary lamp, not looking for the legendary honest man, but "to get @#$%" (sexual satisfaction). I wrote to the editor of the magazine asking for a signed copy. Sure enough, we received a letter with a signed copy of the cartoon—and the usual legal caveats. (I can only assume that the cartoon, which was framed and hung in the DOOM, ended up at AHS when Sinop was closed in 1992.) I have included a photo of the plaque the NCOs gave me when I left in July.

The NCOs' plaque

CSA Gen. Wickham   CSM Welker   Col. Hanne   Col. Stanney   MG Pendelton
U.S. Embassy, Ankara

The "visitors and near-visitors" bit was interesting. One morning, in addition to the daily message from INSCOM and from Seventh Army to "stay alert because there are reports that your facility has been targeted,"[1] there was a message saying that General Wickham, the Army Chief of Staff, would visit Sinop on his tour of the Middle East and Western Europe. Wow! Our little site was on the Chief of Staff's itinerary! This had to be a first! And he was due in shortly, at that. But the closer the day of the visit came, the more tentative the weather became at other locations on his schedule. As a result, CSM Welker and I flew down to the embassy in Ankara to meet with him. We got a nice photo and a thank-you note from the CSA as a memento.

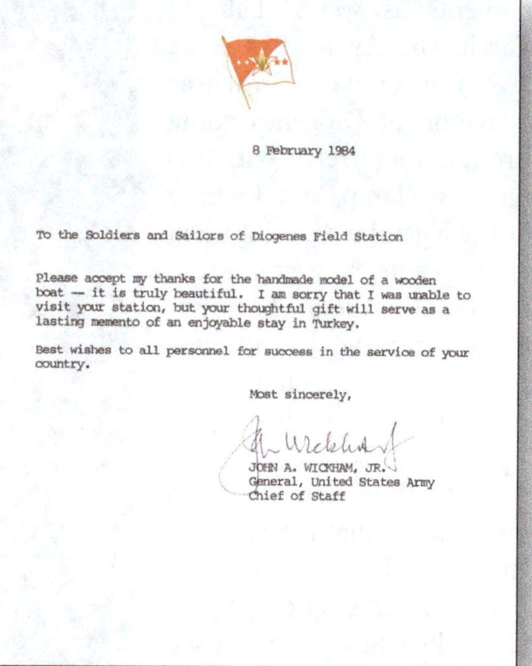

8 February 1984

To the Soldiers and Sailors of Diogenes Field Station

Please accept my thanks for the handmade model of a wooden boat — it is truly beautiful. I am sorry that I was unable to visit your station, but your thoughtful gift will serve as a lasting memento of an enjoyable stay in Turkey.

Best wishes to all personnel for success in the service of your country.

Most sincerely,

JOHN A. WICKHAM, JR.
General, United States Army
Chief of Staff

---

1. Neither headquarters ever said in the almost daily cover-your-ass message who was doing the targeting.

As the end of April drew near, signs of spring were everywhere, especially in the flocks of sheep scattered across the landscape. I had heard of gamboling lambs, and now I could see them: for no apparent reason, a lamb or two would jump straight up into the air and literally bounce across the hillsides. And the museum director kept our heavy equipment and its operators in constant use on the good weekends. We never had any problems, and we could see from week-to-week the improvements that were being made at the museum.

The chaplains planned and carried off a five-day trip to the Holy Lands for interested troops, and there was a sizeable number who participated. Personnel issues were paramount, but the NCOs handled the situation and all positions were covered, in spite of the drawdown in numbers due to the Holy Lands trip. INSCOM operations didn't know about the trip. Of course, I told MG Stubblebine in my periodic report about it, but I didn't think ops needed to know the details, especially after the temper tantrum over earlier shortages.

A quick review of where we thought we stood in our efforts to make our objectives showed progress across the board. We felt vindicated to a large degree because it was about this time that the field station won the INSCOM reenlistment award. We had the highest percentage off eligible personnel reenlist. We took full credit, even though there were other factors in play. The recession of the early 1980s was still in effect in the States, so jobs were hard to come by, plus a troop could reenlist for a station of his or her choice if there were a vacancy at that location. Many of ours who were eligible opted for "current duty assignment."

HIGH PERFORMANCE GRID

	REACTIVE	RESPONSIVE	PROACTIVE	HIGH PERFORMANCE
TIME FRAME	PAST	PRESENT	FUTURE	FLOW
FOCUS	DIFFUSED	OUTPUT	RESULTS	EXCELLENCE
PLANNING	JUSTIFICATION	ACTIVITY	STRATEGY	EVOLUTION
CHANGE MODE	PUNITIVE	ADAPTIVE	PLANNED	PROGRAMMED
MANAGEMENT	FIX BLAME	COORDINATION	ALIGNMENT	NAVIGATION
STRUCTURE	FRAGMENTED	HIERARCHY	MATRIX	NETWORKS
PERSPECTIVE	SELF	TEAM	ORGANIZATION	CULTURE
MOTIVATION	AVOID PAIN	REWARDS	CONTRIBUTION	ACTUALIZATION
DEVELOPMENT	SURVIVAL	COHESION	ATTUNEMENT	TRANSFORMATION
COMMUNICATION	FORCE FEED	FEEDBACK	FEED FORWARD	FEED THROUGH
LEADERSHIP	ENFORCING	COACHING	PURPOSING	EMPOWERING

We believed we were making progress toward our objective and while the rotation of personnel was still an issue, we were getting NCOs and warrant officers who did not appear to have the "I've been exiled" look to them as they got off the plane. Newcomers fit right in, and those who had a previous tour in Sinop appeared to enjoy the sense of purpose-driven activity felt throughout the station.

On a personal level, April was a bummer. An early morning call from Mark Powe at MILPERCEN informed me that I was going to be assigned to Fort Meade. I reminded Mark of our conversation of almost a year ago to the day and of the note I had from Major General Lawrence. Mark said he never promised me that I would return to Carlisle Barracks, and my rebuttal was along the lines of "why bother Lawrence" for agreeing to my return. Essentially, the conversation went nowhere. I was a victim of the Golden Rule: he who has the gold, rules.

I eventually learned that three "bull elephants" (a reference to general officers) were involved in my own assignment and that of one other field-station commander. The Army had not been sending successful field-station commanders to the Agency in the past, but had been keeping them in Army positions, rather than assigning them to the joint slots that the organization wanted. So the Army was "forced" into assigning two of us to the organization to show the Army's support of joint commands. Needless to say, this logic went nowhere with the family in Carlisle. I waited until a trip back around the end of April to break the news in person. I thought that being there in person would help. It did not.

Chapter Ten
Fourth Quarter of the Tour

In late April or early May, all INSCOM commanders were called back to Arlington Hall Station for a command briefing, essentially an update session for the Commanding General, his staff and each other. The biggest advantage was the opportunity to spend some time with Anne and Eric. I was in a losing position with Eric. As a parent, I had long ago made it a matter of record that I would not promise what I could not deliver. Based on what I was told when I took the isolated tour, I thought that what had been promised me—a return to Carlisle Barracks—was money in the bank. I was wrong. I had two choices: 1) take the assignment or 2) retire.

The Army was trying to cut back on personnel, so my retirement would be one less bad-news letter MILPERCEN would have to send. I had 24 years of service, so I could retire as a colonel with a healthy retirement income, but neither Anne nor I were really willing to hang it up. Eric had friends whose parents offered to act as his foster parents in Carlisle, but Anne and I rejected the proposals in favor of mucking it out together, which we ended up doing. It eventually worked out for the best, but there were painful times for all of us along the way.

The INSCOM briefings were almost a repeat of last year's, but I had the chance to lay out what the OE (Organizational Effectiveness) team had done for us, and the progress we had been making, especially in making the assignment a challenging experience for all ranks. I don't think anybody believed me, but I tried.

I flew back to a mostly-new staff. CSM Welker had rotated out, as had Norm, so I had a new command team. Their focus was on the future and what changes Ted (Colonel Ted Fichtl, the commander-designee) would be bringing in with him. He had already selected his operations officer, who would soon be on the scene, replacing Hank Scarangella.

I had a sort of lost feeling for a while. In fact, I learned why my not allowing the display of a short timer's calendar had been a wise decision. On a close-knit team, the rotation out of a member of that team initiated—in some level or degree—the five steps related to the death of a loved one: denial, anger, bartering, depression and finally, acceptance. There are definite stages that the team goes through. Thus, the open display of a double-digit midget (99 or fewer days to go before DEROS—Date Eligible for Return from Overseas) accentuated the inevitable loss to that team and lessened the team ethic.

On the more positive side, our work with the museum director paid off in spades. I was approached, through Octay, with an offer from the director to act as a free tour guide of the area for as many people as would fit in the bus. I wasn't sure if we could turn out enough interested people to make it worthwhile, but we did, and we found a Saturday that worked for all of us.

The Saturday bus tour

The director started us off with some of the earliest-known sites, working the history of the Anatolian Plateau and Black Sea Coast into the history of Turkey, from the Hittites to the end of the Ottoman empire. Our visit to a caravansary[1] was

---

1. An inn surrounding a central courtyard in eastern countries where caravans rest at night.

The remains of Crusader castles and strongholds

The remains of a Crusader fort

especially interesting. It had a built-in heating systems for the stalls or cubicles. The stone walls were built with open ductwork going from the fireplace to a duct in each cubicle, allowing the warmth from the fireplace to reach each little room. We visited the remains of Crusader castles and strongholds, the remains of the forts overlooking Sinop and the bay, and ended the day back at the museum, where we saw displays of amphorae and other items of antiquity collected from nearby sites or the harbor itself. The participants really seemed to enjoy it. I know I did.

As we were saying goodbye in mispronounced Turkish, the director asked if I would be available to spend the next Saturday on a private tour. I jumped at the chance, and it was well worth it! He took me to a secret iron-age dig that Turkish archeologists were working up in the mountains overlooking Sinop.

I took a whole roll of pictures on film—remember, this was 1984—much to the dismay of a couple of workers, but I promised the director that I would not tell anyone about the site. It appeared that every time a historical site was uncovered, the Germans would show up and before the Turks knew it, the artifacts would show up in Berlin or elsewhere. So, this site was off limits to everyone but the Turks. From there, we went on to sites that we had missed the previous Saturday. When we parted, the museum director gave me a small package as a going-away memento of our cooperation with the museum. It contained pieces of amphorae and other remnants of Roman grave sites.

Of over 140 rolls of color film developed by the local photo service in Sinop, only one was "lost." The owners had no idea what had happened to that roll, but they promised that they would look for it. Three guesses as to which roll of film it was, and the first two guesses don't count! That iron age site is still safe. Would I have done anything different if I had been one of the workers? Not on your life! After visiting the museums in Istanbul and Ankara where there are plaques that say, "Replica; original in Berlin (or United States or London)," I fully understand and appreciate their concerns.

The remains of an old Christian chapel on site

The director and I both lamented the gradual decay of an old Christian chapel on our installation. Throughout the year, we had three or more interested troops try their hand at doing something to stop the decay, but it was a bridge too far for us. We were there as part of NATO and not part of an archeological expedition.

———————————————

Weekends were filling up with activities, mainly generated by the NCOs and team leaders. One was the first Sinop "Volks March," a ten-kilometer walk through the region beyond the city of Sinop. The weather was good, and everyone had a good time. I also learned how fast an Anatolian Shepherd (canine, not human) can run. I tried a shortcut and quickly learned that I was trespassing on the shepherd's turf. Ever see those signs that say, "My Doberman can make it to the wall in 3 seconds; can you?" Fortunately, I had not crossed the line yet when the dog came from within the flock of sheep and caught my attention. I can imagine the headlines in *The New York Times*: "Field Station Commander Caught Trespassing; Bite Marks as Evidence."

The first Sinop "Volks March"

The next weekend saw the "Airfield to DOOM Run." This was a 10-kilometer run from the airfield (sea level) to the DOOM (700-foot elevation) through the city of Sinop. This was one event I had to get involved with, as we had women who wanted to run sans excessive attire. Translated, that meant American women running through a conservative Sunni community in shorts and tee-shirts. Men doing so was no problem, but I had to get the mayor's and the governor's okay to do so with the women. Thank God for those countless cups of extra strong Turkish coffee! We got the okay this once, and I assured them I would not be back. But I didn't commit the incoming commander, Ted Fichtl, one way or the other. I came in second behind our surgeon, Bob Snyder.

Three weekends before my change of command we had a unit party to raise funds for Army Emergency Relief, with the big bucks going for the opportunity to bury the old man's face in a cream pie. It turned out that the Navy commander (Clyde Lopez had rotated out, and I do not remember his replacement's name) and my new Deputy, Jimmy Webster, both bid some ridiculous sum for the privilege. It was agreed by them—after all, I'm an innocent pie-stander at this point in the discussion—that one would come from

the right and the other from the left on the count of three. One…two…three…and— I ducked down as they swung their pies…into each other's faces! My timing was perfect. Of course, I got mauled by all after such a stunt, but it was worth it! Then we had the last Hill

Party, as we came to call them during my tour, for the Fourth of July. We invited guests from the community to enjoy an American Fourth of July. It was a great afternoon, with good food and good company, and an opportunity to increase international understanding.

On the non-partying side of the ledger, I used the development of evacuation plans as a training device. We were continually falling short in our planning. I found out in the process that most exposure to planning techniques and procedures were part of on-the-job training so we did some OJT. Over a four-week period we researched, developed, and critiqued plans for the evacuation of the field station. I made it part of a weekly Officers' Call late on Friday afternoons. It was an interesting drill, and the general consensus was that it was a valuable exercise.

Ramadan fell across this period, and Octay invited me to attend noon services one Friday. I did so, but even as a Member of the Book, I was relegated to the balcony of the mosque. Times were simpler then, and many were far more trusting. (Chaplain Thorne was not overjoyed when I told him of my ecumenical activities.)

Earlier snapshots of our progress, or lack thereof, in developing a High Performance Team involved members who had been through at least two earlier snapshots. By late July, we were down to maybe three of us who had spent some time on the activity. Ironically, I found myself lacking enthusiasm to give it another try, but we did one last assessment which is summarized below. There was agreement among the participants that

HIGH PERFORMANCE GRID

	REACTIVE	RESPONSIVE	PROACTIVE	HIGH PERFORMANCE
TIME FRAME	PAST	PRESENT	FUTURE	FLOW
FOCUS	DIFFUSED	OUTPUT	RESULTS	EXCELLENCE
PLANNING	JUSTIFICATION	ACTIVITY	STRATEGY	EVOLUTION
CHANGE MODE	PUNITIVE	ADAPTIVE	PLANNED	PROGRAMMED
MANAGEMENT	FIX BLAME	COORDINATION	ALIGNMENT	NAVIGATION
STRUCTURE	FRAGMENTED	HIERARCHY	MATRIX	NETWORKS
PERSPECTIVE	SELF	TEAM	ORGANIZATION	CULTURE
MOTIVATION	AVOID PAIN	REWARDS	CONTRIBUTION	ACTUALIZATION
DEVELOPMENT	SURVIVAL	COHESION	ATTUNEMENT	TRANSFORMATION
COMMUNICATION	FORCE FEED	FEEDBACK	FEED FORWARD	FEED THROUGH
LEADERSHIP	ENFORCING	COACHING	PURPOSING	EMPOWERING

the concepts of "power down" and "freedom to fail" were worth retaining and adopting. Many of the other concepts, such as "navigation" and "matrix," were a bridge too far and with mission coming first, retraining and follow-through on techniques were often ignored. There was also agreement that MBWA was a lifesaver for the unit, simply because one never knows when the door is going to open, and the boss is going to walk in.

The officers got together under LTC Webster's guidance and threw one hell of a Dining-In two days before the Change of Command. I was the brunt of jabs, jokes, parodies, you name it. The highlight, at least in their eyes, was the receiving line. We

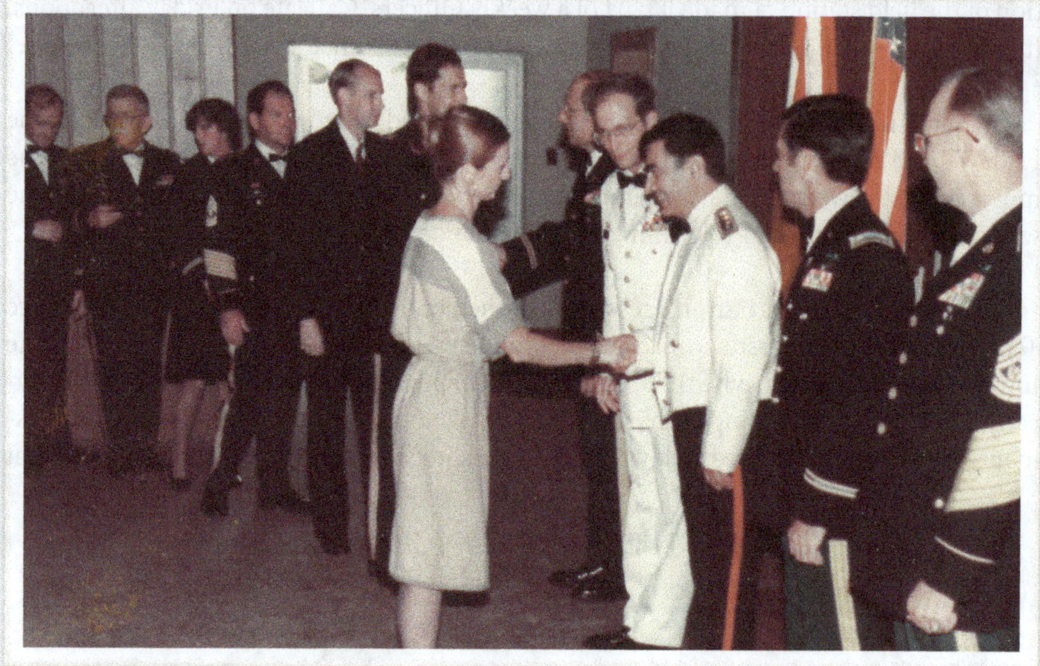

Dining-In two days before the Change of Command

had Colonel Senal and other dignitaries in the line, with me at the end acting as the host. As each officer shook my hand he or she had a wrapped condom concealed in their hand, which they slyly off-loaded into my own hand. I nearly jumped when the first officer came through and shook my hand. I quickly looked to see what it was, although I already had an idea from the feel of it. With the pressure of people coming through the line, I had a real juggling problem on my hands which, of course, elicited all sorts of laughs and snide comments. To me it was the personification of the life in Berlin vs. Vienna comparison that Norm came up with in August a year earlier: The situation here was always hopeless, but never serious!

The Change of Command ceremony two days later was an emotional event. MG Ed Soyster was the new INSCOM commander and a runner as well. He and I had an interesting and informative run the morning of the ceremony. The Mayor was at the ceremony, as were the Governor and the Chief of Police. The ceremony was short, professional, and meaningful. There was no band, as there would have been at a larger command, but we were literally on the front lines in a very stripped-down and lean condition. Hence, we used a tape recorder in lieu of a band, and I provided a recording of *Old Comrades*, an old Prussian march for the Review of Troops.

The Mayor and the Governor gave me the traditional Key to the City, Colonel Senal gave me the Turkish flag that flew over the post on that day, and MG Soyster gave me the American flag that also flew that day. I held my emotions in check right up to the moment when I was driving by the headquarters building on the way out. A contingent of MPs stopped the car and escorted me to the area of the flag pole, at which time they played (on a tape recorder, of course) the National Anthem while they slowly brought the flag down, folded it, and gave it to me. I started to say something, but broke out in tears. It was too much for me to handle. I saluted them, got into the car, and

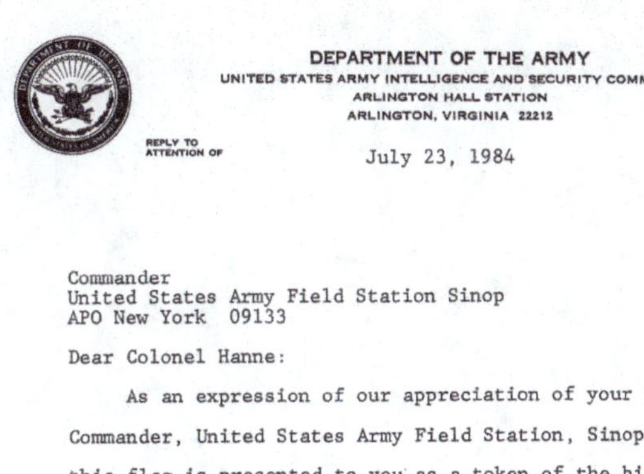

DEPARTMENT OF THE ARMY
UNITED STATES ARMY INTELLIGENCE AND SECURITY COMMAND
ARLINGTON HALL STATION
ARLINGTON, VIRGINIA 22212

REPLY TO
ATTENTION OF

July 23, 1984

Commander
United States Army Field Station Sinop
APO New York 09133

Dear Colonel Hanne:

As an expression of our appreciation of your service as Commander, United States Army Field Station, Sinop, Turkey, this flag is presented to you as a token of the high regard in which your superiors, peers and subordinates hold you. This flag, which flew over your installation the last day of your command, 23 July 1984, is officially retired.

HARRY E. SOYSTER
BG, USA
Commanding

was driven down to the strip where the air crews were assembled to say goodbye. That was another rough parting because it was there that I received the plaque that started this whole story. (See page 3.)

It had been a most rewarding and enjoyable year. The best news came a year later when Field Station Sinop was selected as runner-up for the Travis Trophy equivalent award for installations the size of Sinop. Goal achieved!

Part III

Lessons Learned

Chapter Eleven
Similarities with Peters and Waterman

We had been in command for about four months when Norm burst into my office brandishing a paperback in the air like it was the lost gospels. He proceeded to show me virtually line-by-line what Peters and Waterman had developed in their *In Search of Excellence*,[1] an analysis of high-performing companies. Norm insisted that I must have read the book before taking command, but I had never seen it before. I reminded him that the goals and objectives we had developed in August were derived from our own experience, with the assistance of the OE team, all in house.

Nevertheless, comparing Peters' and Waterman's efforts in their book with what we had done and were in the process of carrying out, was most interesting. They offered eight observations. When comparing theirs against ours, it was apparent that some of their ideas did fit our environment. Their eight observations, in a much-condensed format were the following:

1. Do anything, rather than allow "paralysis through endless analysis."
2. Stay close to your customer; meet his needs, not your preferences.
3. Autonomy—break the organization into bite-size pieces.
4. Encourage buy-in by employees.
5. Stay in touch with your essential business.
6. Stick to your knitting.
7. Have a simple form and a lean staff.
8. Loose/tight relationships: select the key items you need to have tight control over.

---

1. Thomas J. Peters and Robert H. Waterman, Jr. *In Search of Excellence* Published by Harper & Rowe © 1982

We had selected a long term goal: to be highly competitive in the contest for the Travis Trophy. We had two slogans that distilled the essence of our drive. The drive for our goal had a performance side and a "people" side. Our performance slogan stressed a professional attitude (The Best There Is), while the personal slogan stressed working together and looking out for each other (Give a Damn).

Our selection of six separate, but mission-related, objectives reinforced both the goal and the importance of the troops who would achieve that goal. The command structure set the standards and assisted where necessary, while allowing the concepts of "power down" and "freedom to fail" to dominate the setting. The standards were five in number: professional competency, integrity, self-discipline, trust, and communications. These standards were both personal and individual, as well as field-station-wide in application, regardless of one's position or rank.

MBWA assisted in ensuring that a quality product was developed, our first objective. We ensured that from the first day on-site to their departure at the airstrip, the troops knew our goals, the objectives to be taken en route to those goals, and the basic values we held firm. The third objective was enhanced by the remoteness of the site—resources were hard to come by: both human and material resources—so what we had was to be husbanded carefully. The fourth objective, personal and professional development, was enhanced by ensuring that a variety of learning resources were available and used. Our fifth and sixth objectives, to improve the quality of life and build a sense of community, were linked not only together but to all of the other objectives.

In relation to the ideas of Peters and Waterman, we stressed action—immediate action—in lieu of talking it through. Communications in the modern era are often instantaneous, and their content can become meaningless if delayed. Since our customer, the Department of Defense, set the standard and the requirements, they were easy to identify and to maintain. Using "power down" encouraged the development of self-contained, yet interconnected, work sections, allowing for immediate action applied at the right point in a timely manner if such was needed.

First-line supervisors were responsible for the buy-in by their teams, and the NCOs performed in an outstanding manner to the point that there were no outside critics or non-achievers. We made it our first objective to produce a quality product. And we refrained from casting about for other interesting projects, recognizing that our people had been selected for the tasks they performed because of their skills, skills that were not always transferable to other projects without wasting time and effort. The chain of command and of information was simple and lean: "from my lips to their ears," to paraphrase an old Jewish expression. The soldier on the floor had one level to go through to get to me. Sometimes MBWA even eliminated that intermediate level of supervisor. And we practiced the use of loose/tight relationships. Not everything required 24-hour observation, but along with a few other items, the three M's (meals, money, mail) did require it, whether it was at the individual level or field-station level.

Several weeks after I left the Diogenes Field Station, I had the opportunity to read BG James Hunt's and MG Stubblebine's separate evaluations of that year. Hunt was my immediate senior, the Deputy Commander of the U.S. Army Intelligence and Security Command (INSCOM)) and Stubblebine was the INSCOM commanding general. Evaluation reports are done on an annual basis upon completion of an assignment or upon the transfer of your "boss." In this case, all three criteria establishing the need for an evaluation report were met.

BG Hunt summarized my responsibilities as follows:

"Responsible for administration, security, operations, training, programming and budgeting, Facilities Engineer, logistics, and maintenance. Supervises and maintains Base Maintenance Contract. Interfaces with the Country Team, JUSMMAT [Joint U.S. Military Mission Advisory Team] and TUSLOG [Turkey-U.S. Logistics Team]. Senior U.S. official in Sinop Province. Maintains liaison with city, province and martial law authorities. Implements, locally, provisions of the Defense and Economic Cooperative Agreement with Turkey. Coordinates integrated operations and base security with local Turkish commander."

BG Hunt then set forth his opinions as to how these responsibilities were met (skipping the subjective comments and focusing on the objectives attained):

"[He] established an environment of personal growth, high productivity and total involvement by all personnel...Post security was enhanced, quality of life improved and soldiers were challenged to participate in these decisions and the management of their areas of responsibility. They responded. Retention was tops in the MACOM [Major Command, in this instance, INSCOM]. He ensured that government resources were efficiently and effectively managed. Funding obligation achieved a 99.9% rate for both 1983 and 1984. Community involvement and relations with the Turkish military are at an all time level of excellence. A 100% inventory was completed that purged the [stockage] from 7,000 to less than 3,000 lines [of inventory]; and valuable excess equipment was turned in. He obtained a high score on both the Command Maintenance Inspection (CMI) and the Command Supply Inspection and achieved the CMI first pass for this unit in years. Soldiers complaints have become non-existant [sic]. His soldiers are well trained, [proficiency test] scores are high and unit emergency plans are current and tested...[he] has excelled across the

board to make Sinop a productive, efficient, enthusiastic and highly competitive organization."

MG Stubblebine was my senior rater, i.e., my boss's boss and had the final word on the evaluation report. His comments follow. As you read them, you might remember their origin:

> *"Tough minded commander. Demanded much, but not more than he gives. A 'futurist' who provided detailed plans for every step. Surmounted long-standing supply and property accountability problems and fund obligations rate neglected by series of predecessors. Powered down by training, challenging and motivating subordinates. Morale highest. Evidence is: zero complaints, highest MACOM retention rate and three-fold increase in CFC [Combined Federal Campaign]/community involvement. Turkish relations smooth...Brought undesirable tour into a high morale, high spirited, toughly competitive totally professional posture."*

Go back and look at the list of responsibilities, look at the photographs of where elements were located, consider the demographics of the site, look and see the end of the world from any place on the Hill, and you will understand that it was not a one-person show. The successes we had were the result of the actions of the troops I was fortunate enough to have with me, but above all, the successes were the result of the teamwork exemplified by CSM Russ Welker and LTC Norman Chung. Without those two I could not have accomplished what I did.

Because of the caliber of our troops, it was an interesting year—the best of my nearly thirty years of active duty when I retired in 1989. I hope it was the same for those with whom I served at Sinop.

The issues we had identified back in August 1983—fraternization, micro-management, alcohol abuse (essentially a morale issue), and a lack of focus on mission accomplishment coupled with the need to keep the troops focused—had been addressed over time and had been largely successful in being met through the prudent use of MBWA, "power down" and "freedom to fail." We had no AWOL's (personnel absent without leave), we had only one death (a civilian contractor from a heart attack), no suicides, no courts martial, and no incidents of public intoxication even though we had three clubs. And there was a slow but steady decline in private consumption as indicated by a decline in bottle sales.

From what I heard during MBWA and from a number of NCOs, the fraternization issue was no longer a source of scuttlebutt among the lower-ranking troops. The "unofficial newsletter" attempted to make it an issue, but that did not appear to get any traction.

We took first place in re-enlistment, the education center saw an increase in use, personnel involved themselves with the museum and trips offered by the museum, and they made an effort to preserve what we did have on post. Overall morale appeared to be high. My replacement would find issues I had overlooked and bring a different perspective to the command, all for the command's betterment.

Diogenes Field Station—The Best There Is!

Appendix

Biography

A graduate of West Point, Dr. William G. Hanne retired as a Colonel from the United States Army after nearly thirty years of service, during which time he completed two tours of duty in Vietnam and served in both Germany and Turkey. Holding an M.S. in Geography from the University of Illinois and a Ph.D. in Education Policy, Planning, and Administration from the University of Maryland, Dr. Hanne was also Assistant Professor in Geographic Research at West Point. Dr. Hanne currently teaches Russian history and political geography as a volunteer for the Osher Lifelong Learning Institute program under the auspices of the University of Arizona.

Bibliography

Ascherson, Neal. *Black Sea.* New York: Hill & Wang, 1995.

Barnham, Henry D. *Tales of Nasr-ed-Din.* London: Nisbit & Company, 1923.

Burns, J. M. *Leadership.* New York: Harper and Row, 1978.

Hanne, William G. *Perceptions of Presidential Roles of Community College Chief Executive Officers.* Diss. UMI Dissertation Services, 1991.

King, Charles. *The Black Sea: A History.* Oxford: Oxford University Press, 2004.

Nelson, Linda and Frank Burns. "High Performance Organizations: A Framework for Transforming Organizations." *Organization Transformation.* Ed. Adams, John. INSCOM, 1983.

Peters, Thomas J. and Robert H. Waterman, Jr. *In Search of Excellence.* HarperCollins, 2003.

Ryan, William and Walter Pitman. *Noah's Flood.* New York: Simon & Schuster Paperbacks, 1998.

Tumpane, John D. *Scotch and Holy Water.* Lafayette: St. Giles Press, 1981.

Attribution

The following are licensed as noted:

◆ Topographic map of Turkey (derivative).

   Original available at: https://commons.wikimedia.org/wiki/File:Turkey_topo.jpg
   Author: Captain Blood at en:wikipedia.org
   Licensed under the Creative Commons Attribution-Share Alike 3.0 Unported license.

   The map has been created with the Generic Mapping Tools:
   http://gmt.soest.hawaii.edu/ using one or more of these public domain datasets for the relief:
      ETOPO2 (topography/bathymetry): http://www.ngdc.noaa.gov/mgg/global/global.html
      GLOBE (topography): http://www.ngdc.noaa.gov/mgg/topo/gltiles.html
      SRTM (topography): http://www2.jpl.nasa.gov/srtm/

◆ Anatolian Shepherd Dog during the international dogs show in Katowice, Poland.

   Original available at: https://commons.wikimedia.org/w/index.php?title=
   File:Anatolian_2009_pl.jpg&oldid=76414047
   Author: Pleple2000
   Licensed under the terms of the GNU Free Documentation License

◆ Rock-tombs of the Pontic Kings in Amasya - Turkey.

   Original available at: https://commons.wikimedia.org/w/index.php?title=
   File:Kral_Kaya_Mezarları,_Amasya.jpg&oldid=148625695
   Author: Zeynel Cebeci
   Licensed under the terms of the Creative Commons Attribution-Share Alike 4.0 International

Files of the derivatives are available by request from: http://bookservices.us